Geobotany Studies

Basics, Methods and Case Studies

For further volumes:
http://www.springer.com/series/10526

About the Series

The series includes outstanding monographs and collections of papers on a given topic in the following fields: Phytogeography, Phytosociology, Plant Community Ecology, Biocoenology, Vegetation Science, Eco-informatics, Landscape Ecology, Vegetation Mapping, Plant Conservation Biology and Plant Diversity. Contributions are expected to reflect the latest theoretical and methodological developments or to present new applications at large spatial or temporal scales that could reinforce our understanding of ecological processes acting at the phytocoenosis and vegetation landscape level. Case studies based on large data sets are also considered, provided they support habitat classification refinement, plant diversity conservation or vegetation change prediction. Geobotany Studies: Basics, Methods and Case Studies is the successor to Braun-Blanquetia published by the University of Camerino between 1984 and 2011 with cooperation of Station Internationale de Phytosociologie (Bailleul-France) and Dipartimento di Botanica ed Ecologia (Universite' de Camerino - Italia) and under the aegis of Societe' Amicale Francophone de Phytosociologie, Societe' Francaise de Phytosociologie, Rheinold Tuexen Gesellschaft and the Eastern Alpine and Dinaric Society for Vegetation Ecology. This series aims to promote the expansion, evolution and application of the invaluable scientific legacy of the Braun-Blanquetia school.

George Nakhutsrishvili

The Vegetation of Georgia (South Caucasus)

Springer

George Nakhutsrishvili
Ilia State University
Institute of Botany
Tbilisi
Georgia

ISBN 978-3-642-29914-8 ISBN 978-3-642-29915-5 (eBook)
DOI 10.1007/978-3-642-29915-5
Springer Heidelberg New York Dordrecht London

Library of Congress Control Number: 2012944991

1st Edition Nakhutsrishvili, G. (1999) The vegetation of Georgia (Caucasus). Braun-Blanquetia 15:1–74, Camerino, Italy

Printed on acid-free paper

Springer is part of Springer Science+Business Media (www.springer.com)

Preface

Georgia, an ancient Christian country, occupies the central and southwestern parts of the Caucasus – an isthmus between the Black and the Caspian Sea. In the west, it is contiguous to the Black Sea, in the east to the intermountain depression of the River Mtkvari (=Kura) in the Caspian Lowlands (the Transcaucasian Depression), the northern border follows the high mountain tops of the Greater Caucasus, the southern one crosses the Lesser Caucasus (in the eastern part) and separates the Lesser Caucasus from the adjacent mountain ranges of northeastern Asia Minor. Politically, Georgia borders on the Russian Federation (to the north), Azerbaijan (to the east), Turkey (to the southwest) and Armenia (to the southeast). The borders have changed several times during the history of Georgia, especially in the earlier stages of the formation of the Georgian nation.

Kavkasioni, i.e. the Greater Caucasus range, rises up to 5,068 m (Mt. Shkhara, highest mountain within Georgia) and 5,633 m a.s.l. (Mt. Elbrus, in the Russian Federation), i.e. over 1,000 m above the eternal snow-line. Mountain passes at an altitude of 2,200–3,000 m a.s.l. were used as trade-routes from the earliest times. In the mountains of Kavkasioni, permanent settlements are found up to an altitude of 2,500 m a.s.l. Though the Lesser Caucasus mountains reach an altitude of 3,304 m a.s.l., there are no glaciers and no eternal snow.

The natural conditions of the so-called intermontane lowlands of Georgia, consisting of the Colchic Lowlands (adjacent to the Black Sea, in West Georgia) and the foothills between Kavkasioni (in the north) and the Lesser Caucasus (in the southwest), including the Mtkvari Valley (resp. Kura valley, in East Georgia), are favourable for settlement and mixed farming.

In the east, the Colchic Lowlands are surrounded by the Imeretian Elevation which connects the Greater Caucasus with the Lesser Caucasus. The highest part of it, the Surami range, divides the country into two major natural geographical regions: Eastern and Western Georgia, the first having been called Iberia and the latter Colchis by ancient Greek and Roman writers.

Diversity of climate and relief of Georgia has been highly important for its economic development since the early stages of human activities. Numerous rivers (more than 25,000), which promoted intensive agriculture, eliminated the need for irrigation systems, having proved so essential in the economy of the Ancient Eastt (Egypt, Babylon). Besides, these rivers served as trade routes. Due to their rapid current, the rivers were used for transportation of timber (especially in Colchis).

Georgia is rich in various mineral resources. In the mountains of Georgia, copper was extracted, essential for non-ferrous metallurgy, which by that time reached a high level of development. Rich iron deposits prompted the progress of ferrous metallurgy. Magnetic sands along the Black Sea were of particular importance. Rivers, containing gold dust, were known during the Greek Epoch.

Vakhushti Bagrationi, the prominent Georgian historian and geographer of the eighteenth century, distinguished two botanical and agricultural zones within Georgia: the mountainous and the lowland one. Since the Neolithic Revolution, these two zones represented one closely integrated system. The lowlands were characterized by rich harvest of grain crops, vineyards and orchards. In the mountainous areas, the harvest of grain crops was much poorer, and vineyards and orchards were absent.

Georgia covers an area of 69.500 km^2. The population is 5.5 million (two third being Georgians). The Georgians call themselves 'kartveli' (sing.) and 'kartvelebi' (pl.), hence the name of the country in Georgian language is 'Sakartvelo'. The Georgian language belongs to the group of Kartvelian languages, included in the family of Iberian-Caucasian languages. It is suggested that the Georgian language is related to Anterior-Asian languages, the Basque language (the Basque country lies partly in Spain, in the northern part of the Iberain peninsula, close to the Pyrenees). Connections to the Indo-European languages, however, are uncertain. Written Georgian dates back to the third century B.C. (to the reign of king Parnavaz).

Christianity, adopted as the official religion in 337 A.D., promoted the unification of Georgia and the development of written Georgian language.

In ancient times, Georgian tribes were known as skilful farmers, cattle-breeders and metallurgists. Various archaeological discoveries, as well as ancient oriental and Georgian manuscripts testify to an early economical, social, and cultural development of the Georgian people. According to these data, the leading branches of economy of the country must have been: (1) agriculture (field crop cultivation, vine making, horticulture, vegetable growing, etc.); (2) cattle breeding; (3) domestic craft (weaving, woodwork, blacksmith work, textile production, pottery, ceramics, etc.). Georgians developed plough-land cultivation and various forms of irrigation (especially in the mountain areas) aiming to increase crop capacity of natural grasslands (2–3 harvests per season). Of the 16 species of wheat, 11 are represented in Georgia.

Ancient Georgian manuscripts and special medical books ('Karabadini') contain information on various herbs. The first serious scientific information concerning the vegetation of Georgia was given by Vakhushti Bagrationi (eighteenth century). According to several historical sources and notes of foreign travellers, the plants from different parts of Georgia and from other countries were cultivated in the park of Georgian kings in Legvta Khevi (now Botanical Garden of Tbilisi). In the eighteenth century, foreign scientists became interested in the extreme diversity of Georgian vegetation and flora. In 1852, on the basis of rich collections of Caucasian plants, the Caucasian Museum was founded. In the beginning of the twentieth century, all the branches of botanical science (anatomy, physiology, taxonomy, floristics, phytosociology, etc.) started to develop in Georgia. The

Institute of Botany, which belongs to the Academy of Sciences, was opened in 1933. The book *Vegetation of Georgia* by N. Ketskhoveli was published in 1935 and 1960. Long-standing expeditions to different parts of the Caucasus and of Georgia, as well as monographic studies of separate taxa served as a basis for the eight-volume *Flora of Georgia*. In 1971, the first volume of the totally revised second edition of *Flora of Georgia* appeared (16 volumes (1971-2011) have already been published so far). Key to Plants of Adjara by Dmitrieva (1959) and four volumes of the Flora of Abkhazia by Kolakovsky (1980-1986) are also worth mentioning here. Later, it was decided to publish *The Vegetation of Georgia* under the editorship of the present author (Nakhutsrishvili 1990–1991). A brief description of the vegetation cover of Georgia by R. Kvachakidze was published in 2009. The books *Forest Vegetation of Georgia* by A. Dolukhanov and *The History of Flora and Vegetation of Georgia* by I. Shatilova et al. were published in 2010 and 2011, respectively. Until its complete publication, the author of the present book decided to characterize briefly the main types of vegetation of Georgia and to publish it in English. This present publication bears a special purpose: During the 75-year period of isolation of the USSR and of Georgia in particular, it was not so easy for foreign scientists to get adequately acquainted with the interesting plant life of Georgia and to compare the Caucasian flora and plant communities with those of other countries. Today, Georgian botanists are collaborating with scientists from many other countries.

The Number of Plant Taxa

The flora of Georgia comprises about 4,150 species of vascular plants, of which 260 species are endemic, while the flora of the Caucasus comprises 6,350 species of vascular plants, of which 1,600 species are endemic.

Giorgi (Gia) Nakhutsrishvili
Institute of Botany
Ilia State University and Georgia National
Academy of Sciences

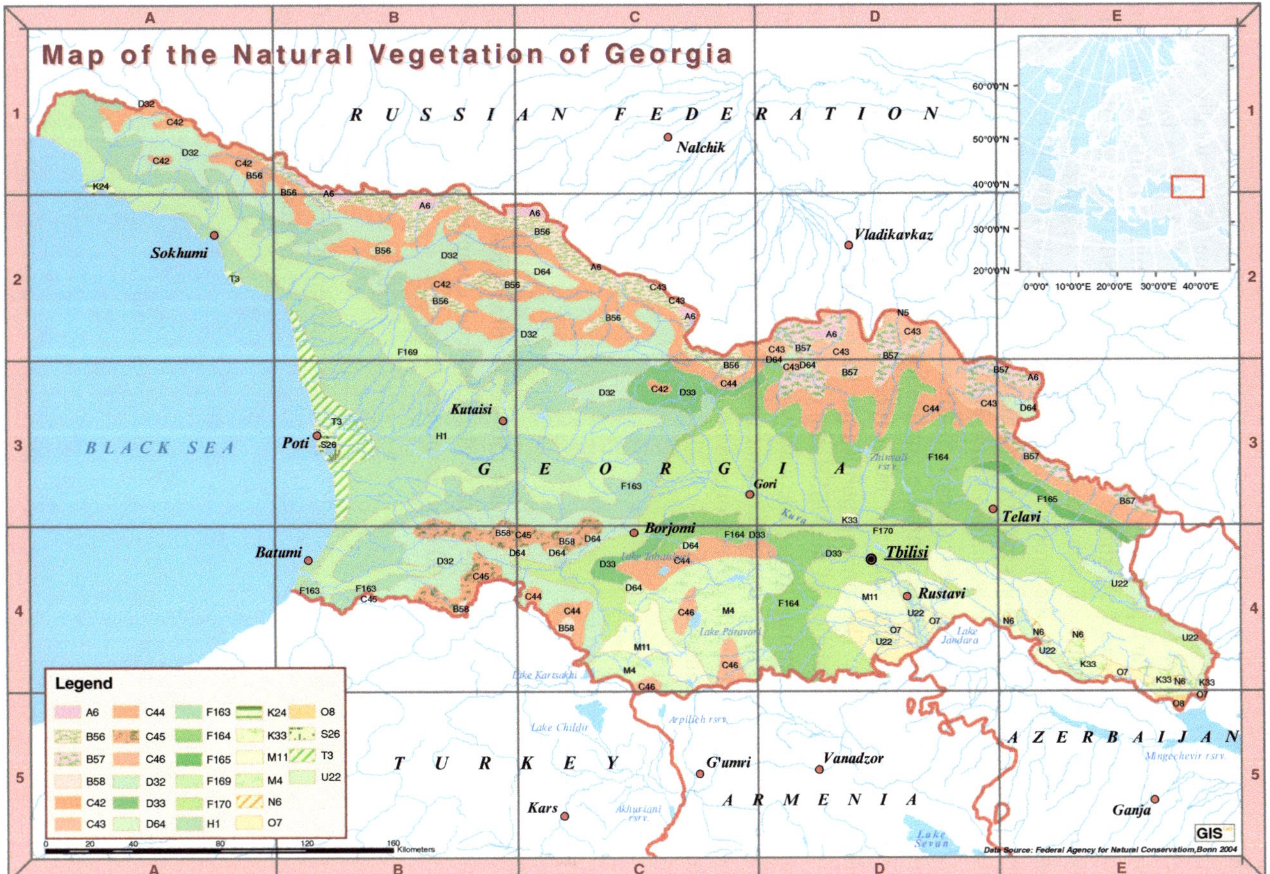
Map of the Natural Vegetation of Georgia
RUSSIAN FEDERATION
GEORGIA
TURKEY
ARMENIA
AZERBAIJAN
BLACK SEA
Nalchik
Vladikavkaz
Sokhumi
Kutaisi
Poti
Batumi
Gori
Borjomi
Telavi
Tbilisi
Rustavi
Kars
G'umri
Vanadzor
Ganja
Kura
Lake Kartsakhi
Lake Childir
Lake Paravani
Lake Tabatskuri
Lake Jandara
Lake Sevan
Zhinvali rsrv.
Arpilich rsrv.
Akhuriani rsrv.
Mingechevir rsrv.
Legend
A6
B56
B57
B58
C42
C43
C44
C45
C46
D32
D33
D64
F163
F164
F165
F169
F170
H1
K24
K33
M11
M4
N6
O7
O8
S26
T3
U22
0 20 40 80 120 160 Kilometers
60°0'0"N
50°0'0"N
40°0'0"N
30°0'0"N
20°0'0"N
0°0'0" 10°0'0"E 20°0'0"E 30°0'0"E 40°0'0"E
GIS
Data Source: Federal Agency for Natural Conservation,Bonn 2004

Expanded Legend

A6 Caucasian open vegetation of lichens and mosses (*Rhizoplaca chrysoleuca, Thamnolia vermicularis, Pohlia elongata, Dicranum elongatum*), with scattered vascular plants on rocky habitats (*Saxifraga ruprechtiana, Primula bayernii*) and on screes (*Cerastium kasbek, Delphinium caucasicum, Symphyoloma graveolens, Scrophularia minima, Lamium tomentosum*)

B56 West Caucasian alpine grasslands (*Geranium gymnocaulon, Nardus stricta, Festuca djimilensis*), calcareous rocks with *Geum speciosum, Carex pontica* and small herb communities (*Sibbaldia semiglabra, Ranunculus brachylobus*), alternating with shrub (*Rhododendron caucasicum*), rock and scree vegetation

B57 East Caucasian alpine grasslands (*Festuca varia* subsp. *woronowii, Nardus stricta, Carex tristis, Kobresia capilliformis*) with *Alchemilla elisabethae* and small herb communities (*Sibbaldia parviflora, Campanula biebersteiniana*), alternating with shrub, dwarf shrub (*Rhododendron caucasicum, Dryas caucasica*), rock and scree vegetation

B58 North Lesser Caucasian alpine grasslands (*Festuca varia* subsp. *woronowii, Nardus stricta, Carex tristis, Bellardiochloa polychroa, Scabiosa caucasica*) with *Agrostis lazica, Bromopsis variegata* and small herb communities (*Carum caucasicum, Campanula tridentata*) with *Gentiana pontica*, alternating with shrub (*Rhododendron caucasicum*), rock and scree vegetation

C42 Southwest Caucasian krummholz and open woodlands (*Betula litwinowii, Fagus orientalis, Acer trautvetteri*) with *Betula megrelica, Quercus pontica*, scrub (*Rhododendron caucasicum*) with Rhamnus imeretina (on carbonate rocks with *Coryllus colchica*), tall-forb communities (*Heracleum ponticum*) with Delphinium pyramidatum (on carbonate rocks with *Heracleum aconitifolium, Ligusticum arafoe*) and grasslands (*Calamagrostis arundinacea, Betonica macrantha*, on carbonate rocks with *Woronowia speciosa, Carex pontica*)

C43 Northeast Caucasian krummholz and open woodlands (*Betula litwinowii, Acer trautvetteri, partly Pinus kochiana*) with *Betula raddeana*, scrub (*Rhododendron caucasicum, Juniperus communis* subsp. *hemisphaerica*), tall-forb communities (*Heracleum sosnowskyi, Delphinium flexuosum*) and grasslands (*Festuca woronowii, Bromopsis variegata*), alternating with dry grasslands (*Festuca ovina, Carex humilis*)

C44 Southeast and Lesser Caucasian krummholz and open woodlands (*Betula litwinowii, Acer trautvetteri, Quercus macranthera*), scrub (*Rhododendron caucasicum*), tall-forb communities (*Heracleum sosnowskyi, Aconitum orientale*) and grasslands (*Festuca varia* subsp. *woronowii, Calamagrostis arundinacea, Geranium ibericum*)

C45 West Lesser Caucasian krummholz and open woodlands (*Betula litwinowii, Fagus orientalis, Acer trautvetteri*) with *Quercus pontica, Betula medwedewii*, scrub (*Rhododendron caucasicum*) with *Rhododendron ungernii*, tall-forb communities (*Ligusticum alatum, Milium schmidtianum*) with *Heracleum cyclocarpum, H. mantegazzianum* and grasslands (*Agrostis planifolia, Geranium platypetalum*) with *Euphorbia oblongifolia, Astragalus bachmarensis*)

C46 East Lesser Caucasian krummholz and open woodlands (*Quercus macranthera, Acer trautvetteri, Betula litwinowii*), scrub (*Juniperus communis* subsp. *hemisphaerica*, J. sabina), grasslands (*Festuca varia* subsp. *woronowii, Bromopsis variegata, Anemonastrum fasciculatum*) with *Trifolium bordzilowsky*, partly tall-forb communities (*Milium effusum, Gagea orientalis*) alternating with dry grasslands (*Festuca ovina, Carex humilis*) with true steppes (*Festuca valesiaca, Stipa tirsa, S. pulcherrima*), partly with thorn-cushion mountain vegetation (*Astragalus aureus*)

D32 West Caucasian fir, spruce-fir and beech-fir forests (*Abies nordmanniana, Picea orientalis, Fagus orientalis*) with evergreen understorey *Rhododendron ponticum, Prunus laurocerasus, Ilex colchica*), often alternating with Oriental beech forests (*Fagus orientalis*)

D33 Caucasian fir, spruce-fir and beech-fir forests (*Abies nordmanniana, Picea orientalis, Fagus orientalis*) without evergreen understorey, partly alternating with Oriental beech forests (*Fagus orientalis*)

D64 Caucasian pine forests (*Pinus kochiana*), partly alternating with birch forests (*Betula litwinowii*, *B. raddeana*) and spruce forests (*Picea orientalis*)

F163 East Euxinian-Caucasian Oriental beech forests (*Fagus orientalis*) partly with *Picea orientalis*, mostly with evergreen understorey (*Prunus laurocerasus*, *Rhododendron ponticum*, *Daphne pontica*) with *Hedera colchica*, *Ilex colchica*, *Ruscus colchicus*

F164 Caucasian Oriental beech forests (*Fagus orientalis*) with *Carpinus C. caucasica* partly with *Picea orientalis*, without evergreen understorey partly alternating with oak-hornbeam forests (*Carpinus betulus*, *Quercus iberica*)

F165 East Caucasian submontane to montane hornbeam-maple-Oriental beech forests (*Fagus sylvatica subsp. orientalis*, *Acer velutinum*, *Carpinus caucasica*) with *Hedera pastuchowii* in combination with hornbeam-chestnut-oak forests (*Quercus iberica*, *Castanea sativa*, *Carpinus caucasica*)

F169 East Euxinian oak and hornbeam-oak forests (*Quercus iberica*, *Carpinus orientalis*, *C. caucasica*), alternating with hornbeam-chestnut-Oriental beech forests (*Carpinus caucasica*, *Fagus orientalis*, *Castanea sativa*) with evergreen understorey

F170 Transcaucasian oak forests (*Quercus iberica*), hornbeam-oak forests (*Quercus iberica*, *Carpinus caucasica*) and Oriental hornbeam-oak forests (*Quercus iberica*, *Carpinus orientalis*), with *Sorbus torminalis*, partly in combination with shibliak communities (scrub)

H1 Colchic lowland to submontane mixed oak forests (*Quercus imeretina*, *Q. hartwissiana*, *Zelkova carpinifolia*, *Carpinus caucasica*, *Castanea sativa*, *Fagus orientalis*) with evergreen understorey species (*Rhododendron ponticum*, *Prunus laurocerasus*), alternating with oak and hornbeam-oak forests (*Quercus iberica*, *Carpinus betulus*) in the submontane belt

K24 West Caucasian *Pinus pityusa*-forests with *Carpinus orientalis*, *Cistus crecitus*, *Ruscus aculeatus*

K33 Transcaucasian colline-montane juniper open woodlands (*Juniperus polycarpos*, *J. foetidissima*), partly in combination with *Pistacia mutica* – open woodlands

M4 Transcaucasian altimontane herb-grass- and meadow steppes (*Stipa tirsa*, *S. pulcherrima*, *Festuca ovina*, *Carex humilis*, *Poa densa*, *Bromopsis variegata*, *Onobrychis altissima*, *O. transcaucasica*, *Aster ibericus*, *Scutellaria orientalis*)

M11 Pre- and Transcaucasian Stipa-steppes (*Stipa tirsa*, *S. pulcherrima*, *S. pontica*) with Onobrychis transcaucasica, Botriochloa ischaemum – steppes with *Onobrychis kachetica*, *Medicago coerulea*, *Polygala transcaucasica*, alternating with tomillares (*Thymus tiflisiensis*, *Scutellaria orientalis*) and thorn-cushion communities (*Astragalus denudatus*, *A. microcephalus*)

N6 East Transcaucasian thorn-cushion vegetation (*Astragalus caucasicus*, *A. microcephalus*, *Acantholimon lepturoides*. *A. fominii*) and tomillares (*Salvia garedji*, *Thymus tiflisiensis*)

O7 East Transcaucasian wormwood deserts (*Artemisia lerchiana*) with ephemeroids (*Poa bulbosa*, *Catabrosella humilis*)

08 East Transcaucasian *Salsola nodulosa* and *Salsola ericoides* – deserts with ephemeroids (*Poa bulbosa*, *Catabrosella humilis*), with *Artemisia lerchiana*

S26 Colchic herb-rich tall sedge fens with *Carex acuta*, *Cladium mariscus*, *Ludwigia palustris* in combination with Sphagnum-mires (*Sphagnum austinii*, *S. papillosum*) with *Rhododendron luteum*, *Osmunda regalis*, *Rhynchospora caucasica*

T3 Colchic alder carrs (*Alnus barbata*) in combination with alluvial forests (*Alnus barbata*, *Fraxinus excelsior*, *Pterocarya pterocarpa*), tall reed vegetation (*Phragmites australis*, *Typha latifolia*) and sedge swamps (*Carex* spp.)

U22 Transcaucasicum hardwood alluvial forests (*Quercus pedunculiflora*, *Ulmus minor*) in combination with poplar and willow alluvial forests (*Populus x canescens*, *P. nigra*, *Salix excelsa*) as well as *Tamarix ramosissima*-scrub

Source: The map is given according to Bohn et al. (2003), where the Caucasus part was prepared by D Bedoshvili, A Dolukhanov, M Ivanishvili, G Nakhutsrishvili, N Zazanashvili.

Acknowledgements

I am very grateful to Mr. David Kikodze, Institute of Botany, Ilia State University, Tbilisi, for translation. Special thanks to Prof. Manfred A. Fischer, University of Vienna, Austria, Prof. Otar Abdaladze, Prof. Maia Akhalkatsi, and Asst. Prof. Ketevan Batsatsashvili, Ilia State University, Tbilisi, who critically revised the manuscript and particularly to Prof. David D. Ferguson, University of Vienna, Austria, who was very kind to revise the palaeobotanical chapter. I thank Dr. Nikoloz Lachashvili for detailed revision of the arid vegetation chapter. Thanks to Mr. Giorgi Mikeladze for vegetation mapping. Special thanks likewise to the Agency of Protected Areas of Georgia for the map of Georgia's protected areas. Thanks to Mr. Giorgi Mikeladze, a PHD student of Ilia State University, for vegetation mapping. Special thanks likewise to the Agency of Protected Areas of Georgia for the map of Georgia's protected areas. I am grateful to Dr. Shalva Sikharulidze, Director, Institute of Botany, Ilia State University, Tbilisi, for support of the publication. Also, I thank Mr. Irakli Rostomashvili and Mr. Zezva Asanidze, a PHD student of Ilia State University, for computer service.

I am especially grateful to Dr. Franco Pedrotti, University of Camerino, Italy, who was the initiator of the publication of this book.

Contents

1 Environmental Conditions

Georgia occupies an interesting geobotanical position as a part of Caucasia – the region which links Europe with Asia. The country is characterized by rather contrasting natural conditions which account for the extremely high degree of divergence of plant communities within this comparatively small area. The landscape of the country includes different types of desert and semi-desert vegetation mainly in the eastern parts of Georgia, luxuriant Colchic forests of moist, almost subtropical climate in the west, and high-mountain plant communities in the north and in the south (See "Map of the Natural Vegetation of Georgia" 2003).

The complex orographic structure of Georgia and its geographical position account for the geographical and ecological isolation of certain plant communities, which has resulted in a high ratio of local endemism (particularly endemics of the Greater Caucasus Range and of the Colchis) as well as a variety of species of different phytochorological groups, like Iberian and Anterior-Asian group, etc.

1.1 Orography

Geologically, the territory of Georgia belongs to the Alpine System of Eurasia. Its geological and geomorphological structure reveals a great genetic diversity, as a result of tectonic, volcanic, petrological, gravitational, erosional and other processes.

Structurally, the area can be divided into the following major landforms:

1. The range of the Greater Caucasus (Kavkasioni);
2. The Georgian Intermontane Area (between the Greater and the Lesser Caucasus);
3. The Mountain System of the Lesser Caucasus (Meskheti-Trialetian Range), including the South Georgian Volcanic Upland.

Georgia comprises the southern side of the middle part of the Kavkasioni as well as the north-western parts of the Transcaucasian Depression, the Lesser Caucasus, i.e. the South Caucasian Upland.

Altitudes in Georgia vary considerably from the sea level (at the Black Sea) to some of the highest peaks of the Greater Caucasus, reaching 4,695 m (mt. Ushba),

G. Nakhutsrishvili, *The Vegetation of Georgia (South Caucasus)*, Geobotany Studies, DOI 10.1007/978-3-642-29915-5_1, © Springer-Verlag Berlin Heidelberg 2013

5,068 m (in the Shkhara massif), and 5,047 m a.s.l. (mt. Kazbegi, Kazbek). From the geological point of view, this area consists, mainly, of Meso- to Cenozoic deposits. Ancient Precambrian and Paleozoic formations are poorly represented and of secondary importance (Gerasimov 1966).

1.2 Climate

The climate is one of extremes. It is possible to distinguish several climatic zones from the humid, almost subtropical climate to the climate of permanent snows and glaciers. Such a considerable range of climatic conditions is caused both by the orographic structure and the presence of the Black Sea and the Caspian Sea. The Black Sea and high mountain ranges of the Greater Caucasus are the most important orographic factors determining the climate of Georgia by preventing the invasion of cold air masses from the north. The climate of the country has been formed by the air masses blowing from sea, as well as by the latitudinal position between southern Russia and the Inner-Anatolian mountains system.

The highest mean annual temperature amounts to 15 °C (Sukhumi in Abkhazeti, on the Black Sea coast). The warmest winter (5–7 °C, in January) is in Colchis (West Georgia, on the Black Sea coast).

Annual precipitations in Georgia range from over 4,500 to 400 mm or less. Increase of altitude e.g. in Svaneti (north-western Georgia) and Javakheti (south-western Georgia) is paralleled by reduction of precipitation, while in other districts (Kazbegi region, Kakheti, etc.) the amount of rainfall increases with elevation. The highest amounts of precipitation occur at the following altitudes: from 300–500 m to 3,500 m a.s.l. in Western Georgia, and from 1,200 to 3,500 m in the eastern part of the country.

The climatic conditions of several districts of Georgia are demonstrated diagrammatically (Fig. 1.1). Notwithstanding the short distance between these districts (60 km between Batumi and Bakhmaro, 380 km between Batumi and Shiraki, 155 km between Shiraki and Kazbegi), their climatic conditions are noticeably different.

1.3 Soils

There is a great diversity of soil types in Georgia. The following soil provinces can be distinguished:

1. The soil province of Western Georgia;
2. The soil province of Eastern Georgia;
3. The soil province of Southern Georgia.

The main types of soil are allocated in accordance with the altitudinal zones. In the Western Georgia soil province (between the Black Sea and Likhi mts.) it is possible to distinguish the bog and podzolic soil zone in the lowlands, the krasnozem and zheltozem zone in the hilly piedmonts, the zone of mountain-forest and the mountain-meadow soils.

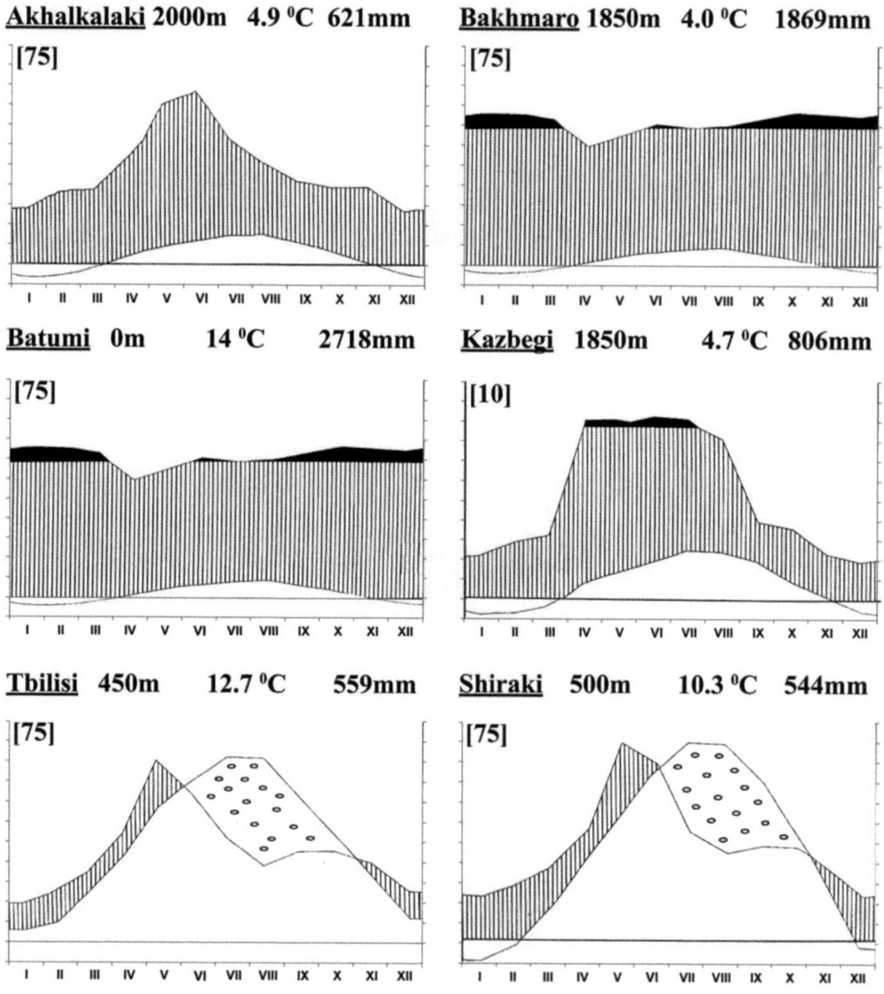

Fig. 1.1 Climadiagrams of several regions of Georgia

The soil province of Eastern Georgia comprises the plains, piedmonts and mountain massifs, situated eastwards from Likhi mountain range. Chestnut soils, chernozems, brown humic-sulphates, saline soils of steppes and semi-deserts, as well as intermediate forest-steppe and mountain-meadow soils occur in this province.

The Southern Georgia soil province includes Javakheti, Tsalka-Dmanisi and Erusheti uplands, the hollow of Akhaltsikhe, etc. A considerable part of this area is covered both with the mountain chernozems (which are formed at altitudes from 1,200 to 2,200 m) and meadow chernozem-like soils. In highlands they are replaced by mountain-meadow soils. Besides, the alluvial soils, redzinas, brown as well as the meadow-brown soils occur here, with the predominance of brown forest type of soil in the mountain forest belt (Urushadze 1987).

History of the Flora and the Vegetation 2

The history of the flora and, especially, the vegetation of Georgia is very complicated and, despite the intensive paleobotanical investigations of the recent past, still needs to be studied.

The earliest records of fossil flora stem from the Paleozoic. Species of *Lepidodendron* (large trees reaching a height of nearly 40 m), *Sigillaria* (high plants with erect stems and characteristic rhizophore system: stigmarias), *Asterocalamites* were determined from the Lower and Middle Carboniferous deposition of the Khrami crystalline massif (South Georgia). Cordaitales – a group with anatomical structures (tracheids, well-developed pith) very similar to that of conifers (Shatilova et al. 2011) – are known from the same period.

In the Early Jurassic, almost the entire territory of Georgia was covered by sea, except for Khrami, Loki and Dzirula ancient massifs.

The climate of this period was warm-temperate. Water temperature in the sea reached 23–24 °C, but in Late Toarcian it decreased considerably (7–15 °C). Calamitales are of special interest with the only species *Neocalamites hoerensis* present in the early Jurassic depositions. The above mentioned period was characterized by the wide-spread occurrence of equisetoids, which were restricted to marshes as it is nowadays (Svanidze 1972).

Pteridophytes were represented by 25 taxa. Microfossils found belong mainly to the genus *Cladophlebis* (Osmundaceae). At the same time *Ginkgo* was represented by two species: *G. mziae* (specific to Georgia) and *G. huttonii*. The species *Sphlenobaiera spectabilis* and *Phoenicopsis angustifolia* also belong to Ginkgoaceae. In the early Jurassic, a considerable group was formed by Bennettitales – the ancestors of the flowering plants (according to the viewpoint of several botanists). *Eucommiidites troedssonii* (gymnosperms pollen) and many other plants were present in the same period too.

In the Bajocian (Middle Jurassic) the entire territory of Georgia was under the sea; by the end of it, first the eastern part and afterwards (in Bithonian) the whole territory has emerged. The appearance of freshwater basins was characteristic to that period, as well as the process of peat accumulation (Kakhadze 1947). The Early Jurassic flora is much richer than that of the Early Jurassic. Today nearly 175 plant

G. Nakhutsrishvili, *The Vegetation of Georgia (South Caucasus)*, Geobotany Studies, DOI 10.1007/978-3-642-29915-5_2, © Springer-Verlag Berlin Heidelberg 2013

species are known from the Middle Jurassic. 55 of them are ferns (*Coniopteris*, Cyatheaceae, Osmundaceae, Gleicheniaceae) and representatives of *Paracycas* (Cycadaceae), *Ptilophyllum* (Benettitaceae), etc. The climate of the Middle Jurassic was probably tropical; representatives of Cycadales and Ginkgoales dominated the woodlands (Svanidze 1972).

In the Late Jurassic the territory of Georgia was covered by sea, except the Svaneti Elevation, as well as the southern parts.

The occurrence of remnants of the new species *Angiopteris iberica* (Marattiaceae) in the Late Jurassic fossiliferous deposits is of great importance. It underlies the opinion that these archaic living ferns disappeared from the floras of the Northern Hemisphere after the Middle Jurassic.

According to Vakhrameev (cited by Shatilova et al. 2011), the Late Jurassic paleofloristic province of the Caucasus was a part of the Euro-Sinian region.

The flora of the above-mentioned province could be characterized by the presence of occasional equisetoids, a few ferns, Ginkgoales, the absence of *Czekanowskia* and the predominance of Cycadales, Bennettitales as well as *Pachypteris*, *Sagenopteris*, *Pagiophyllum*.

2.1 The Tertiary

In the **Paleocene** and **Eocene** a long island was formed on the territory of the Caucasus as a basis for further development of mountain range. In the Eocene, the dominant position was occupied by the Angiosperms, representatives of evergreen Fagaceae and *Myrica*.

At the same time the process of migration of the Boreal cold-resistant plants, which had begun previously in the Cretaceous, resumed at the end of the Eocene.

At the Eocene/Oligocene boundary, the folded structure of the Alpine system started to form in the Mediterranean orogenetic band. It comprised the (geosynclinal) regions of the Southern Slope of the Greater Caucasus range and Ajara-Trialeti, the main part of today's Georgia. Uplift caused a change in the climatic conditions; the latter determined the character of the vegetation: the number of conifers increased and warm-temperate plants became more numerous (Tumajanov 1955; Shatilova et al. 2011).

During the **Oligocene**, the Caucasus was a small island surrounded by a vast basin with scattered islands, where the tropical flora of the Poltava type (palms, laurels, etc.) was well developed. In West Georgia (near the town Chiatura), in the Oligocene deposits remnants of conifers (species of *Pinus*) were found, indicating the penetration of northern taxa into the Caucasus.

The Oligocene plant world was greatly influenced by the Arcto-Tertiary or Turgay flora, formed in the eastern and northern parts of modern Asia. Representatives of this flora, especially pines, penetrated into the Caucasus. Many botanists consider that migration of the Turgay elements became more intensive after the Turgay Strait had become dry by the end of the Oligocene.

According to Grossheim (1948), Turgay migrants began to change and gave rise to the ancient mesophytic forest flora, called Mediterranean-Turgay flora by this author.

During the **Miocene** the Caucasian island started to expand. Fossil assemblages from different parts of Georgia, especially from Western Georgia (Guria), confirm the subtropical nature of the Miocene flora with the predominance of evergreen woody plants, accompanied by the deciduous *Castanea sativa*, as well as by conifers of northern origin *(Pinus neptuni)*.

The following plants have been known since the Early Miocene deposites of Kartli (East Georgia): *Myrica neriifolia* (=*M. lignitum*), *Laurophyllum primigenium*, species of *Apocynophyllum*. Of the 46 taxa two are pteridophytes and five gymnosperms. The dominant position is occupied by Angiosperms, namely *Comptonia acutiloba*, *Myrica neriifolia* (=*M. lignitum*), etc. The greatest part of the Early Miocene flora, studied by macrofossils, is represented by Angiosperms. Among them the predominance of Myricaceae, Juglandaceae and Lauraceae should be noted, whose role has somewhat increased since the Paleogene. Evergreen Fagaceae, Lauraceae and other thermophilous plants, contributing to the formation of moist-subtropical forests in the Paleogene, adapted to the reduced humidity of the Early Miocene.

In the Middle Miocene plants of warm and dry climate participated in sclerophyllous formations. *Myrica* species covered river banks. Sclerophyllous plants were represented by the species of *Myrtus, Callistemophyllum, Acacia, Quercus (Q. drymeja)*. Several species of Lauraceae, *Sapindus, Smilax* and *Magnolia* formed moist-subtropical forests.

It is possible to suggest that an altitudinal differentiation of vegetation already existed in the Middle Miocene. Coastal and low mountainous areas were covered by subtropical forests with high numbers of Sterculiaceae, Araliaceae, Lauraceae, evergreen Fagaceae, arborescent ferns, *Hymenophyllum*, etc. The next altitudinal belt was occupied by mesophilous deciduous forests of *Platanus, Comptonia, Juglans, Pterocarya*, etc. Cold-resistant plants occurred even higher; mention should be made of species of *Betula, Fagus, Acer, Tilia, Ulmus*, accompanied by *Cathaya, Keteleeria*, etc. (Shatilova et al. 2011).

During the Sarmatian (Upper Miocene), the Caucasian island, known in the literature as Jephethis, was situated close to Iranian and Central Asian land. This encouraged the immigration of xerophilous elements into the Caucasus. It is suggested that the forestless formations of xerophytic plant communities were widely distributed during the Early Miocene. According to Grossheim (1946), *Pelargonium endlicherianum*, the fern *Ceterach*, etc., are relicts of the ancient xerophytic flora of this period.

The close position of Jephethis to the Iranian mainland, separated as a cape, caused a division of the mesophytic Mediterranean-Turgay flora into western and eastern parts. In the west, the Colchic flora began to form, while the Hyrcanian elements evolved in the eastern part. Both were of mesophytic nature. The analysis of the contemporary vegetation makes many botanists suggest the existence of a third xerophytic flora.

Based on the fossils, it is possible to build up a picture of the Sarmatian flora: evergreen *Magnolia dianae*, species of Lauraceae (*Cinnamomum, Laurophyllum primigenium, Laurophyllum pseudoprinceps*, etc.), as well as the representatives of the Mediteranean-Turgay and Turgay deciduous flora, such as *Salix, Pyrus, Carpinus, Juglans, Ulmus*, etc. Conifers (*Pinus saturi, Sequoia langsdorfii*) and xerophytic species of *Myrica* have been known from the same age.

Kolakovsky and Shakril (1976) pointed out that the richest Sarmatian floras occurred in Abkhazeti (Colchis). They were closely related to the southern subtropical and tropical floras (*Melastomites* sp., *Mastixia microphylla, Ventilago* etc.). The Sarmatian floras of Abkhazeti are very similar to Oligocene-Miocene European floras. These floras indicate the wide distribution of subtropical evergreen plants, mainly Lauraceae, such as *Ocotea, Persea, Aniba*, and many others, and the deciduous arborescent plants of warm climate which have survived until now only in Central and South America, partly in the Mediterranean area and in South-Eastern Asia. These forests show a great similarity to the vegetation, confined to mountainous Japan during the Miocene-Pliocene. It is suggested that this vegetation is similar to that now found in the mountain systems of South-Eastern Asia, Himalaya and China, where relic forests of *Ulmus, Fagus, Quercus, Abies* and *Cryptomeria* were well represented.

The Tertiary-relic species of Colchic type, such as species of *Buxus, Pterocarya*, etc. and the hemixerophytic Mediterranean plants *Arbutus, Celtis, Smilax, Thelycrania (Cornus), Quercus*, have been found in the Sarmatian deposites of Abkhazeti.

The territory of East Georgia, where steppes and semi-deserts are distributed now, is considered to have been covered with both sclerophyllous and moist-subtropical forests.

In South Georgia the shrubs of subtropical and temperate climate dominated by Lauraceae, were represented by narrow-leaved xerophytic elements.

In the Meotian, after the regression of the Mediterranean Sea, the strait separating the Jepethis from the southern mainland disappeared and the Caucasus became a peninsula. The formation of the main folded systems reached the final point, and the vegetational belts were formed. The Meotian macrofossils included *Cryptomeria japonica*, species of Lauraceae, *Salix coriacea, Myrica neriifolia* (=*M. lignitum*). Ferns, especially *Dicksonia*, were represented by a large number of species. Palynological data attest to the fact that the following genera of conifers were present: *Abies, Picea, Tsuga* (five species), *Cedrus*, as well as *Carya, "Castanopsis furcinervis"* (leaf remnants of *Quercus*), representatives of Hamamelidaceae or Altingiaceae (*Liquidambar*, etc.).

According to the composition of the fossiliferous deposits, in this period, coniferous and broad-leaved forests (*Taxodium, Cryptomeria, Magnolia*, species of Hamamelidaceae, etc.) with the species of *Quercus, Carya, Fagus, Castanea* and *Platanus* began to prevail.

Relatively dry habitats were occupied by the following hemixerophytes: *Quercus drymeja, Celtis punica, Pistacia miocenica, Sophora europaea, Pyracantha coccinea*. At the same time the following plants began to crop up (present-day members of the Georgian flora): *Rhododendron caucasicum, Fagus orientalis, Zelkova crenata (=Z. carpinifolia)*. During the period of Pontian Sea expansion, the Caucasus and Transcaucasia became consolidated.

The fossil flora of Goderdzi pass (South-Western Georgia) was distinguished by the features characteristic of Sarmatian flora; in Pontic it had already been a relic flora.

Taking into account fossil deposits of this region, mesophytic flora obviously prevailed here. Representatives of Annonaceae, Hamamelidaceae and Lauraceae are known from this period which was characterized by a warm maritime climate. Ferns, distributed in Georgia during the Pontian, can now be found only in the tropics.

During the Pontian, gymnosperms were widespread. Nowadays, they occur in montane forests of tropical and subtropical countries. In the beginning of the Pliocene, *Pinus pithyusa* was widely distributed.

Investigations carried out by Kolakovsky (1964) in Kodori (Abkhazeti) point to the existence of lowland and low-mountain forests with *Myrica lignitum, Salix varians, Alnus subcordata*. Coastal lowlands were covered with forests of *Quercus kodorica* and *Carya denticulata*, while the riversides were occupied by species of *Liquidambar, Taxodium, Nyssa, Tectocarya, Ocotea,* evergreen Fagaceae like *Pasania, Castanopsis*, etc., which were characteristic of the subtropical forests of the Pontian. In the lower montane zone, considerable areas were populated by Laurisilva with tree ferns *(Cyathea, Alsophila, Dicksonia).*

In the lower montane zone sclerophyllous formations were distributed with the elements of maquis. Pines occupied rocky habitats.

Kolakovsky (1974) pointed out that an important change in the flora of the warm-temperate climate took place during the Pontian. The mid-mountain belt was populated by *Fagus, Castanea, Acer, Tilia* species. The presence of conifers – *Ginkgo, Keteleeria* – , ferns like *Polypodium*, etc., lianas like *Vitis betulifolia, Parthenocissus quinquefolia* becomes fairly possible.

Dark coniferous forests (with species of *Abies, Tsuga, Picea, Cedrus,* etc.) occupied relatively higher altitudes.

On the territory of West Georgia a great number of angiosperms has vanished since the Pontian.

In Cimmeridian (Middle Pliocene), the shrinking of the Tethys led to the exposure of new land, fit for colonization by xerophytic vegetation, such as semi-desert vegetation, shibliak, light forests, maquis, etc., so characteristic of the Mediterranean. Meanwhile, taxa of Araliaceae, as well as ferns were well represented. The dominant position was occupied by species of *Pteris* and *Polypodium*. According to Kolakovsky and Shakril (1978), forests of warm-temperate climate began to prevail in Cimmeridian; they were composed by the species of moist monsoon climate, now peculiar of the mountains of Eastern Asia and xerophytes, ecologically similar to the plants of xerophytic forests of the Mediterranean. Above the warm-temperate forest zone hardwood and coniferous forests grew.

During the Kuyalnitskian-Akchagylian (Upper Pliocene), a large part of East Georgia lay under the sea. The Greater and Lesser Caucasus were connected by a land bridge to the Near East (Asia Anterior–South-West Asia) in the south. The most striking feature of the fossil floras of this period is the absence of Poltava floral elements and evergreen plants. The following plants are known from this period: species of *Populus, Ostrya, Fagus, Zelkova, Tilia, Pyrus, Ligustrum, Salix, Prunus, Acer, Quercus, Pterocarya*, together with representatives of *Carex* and *Phragmites*. All these plants are members of the present-day flora of Georgia, though they have changed their habitats. For example, remains of *Fagus orientalis* and *Ostrya carpinifolia* have been found in Akchagyl deposits of the Shiraki steppe. Due to increasing aridity, these trees are no longer there.

According to Grossheim (1948) the Akchagylian flora by its nature is closely connected with that of the Quaternary. To quote his words: "The period between Sarmatian and Akchagylian was the turning-point in the floristic history of the Caucasus; the tropical flora had been replaced by that of temperate climate" (Grossheim 1948: 171). Broad-leaved forests were replaced by conifers, namely species of *Picea, Abies,* and *Tsuga*. In general, during the late Pliocene, the role of polydominant forest vegetation was reduced and, instead, the formation of communities with one or two dominant species was stimulated. These changes in the vegetational cover were probably provoked by the intensification of cold at the end of the Miocene and in the Pliocene.

The existence of a dry and hot climate between the cold periods encouraged the enrichment of the Caucasian flora with xerophytes. In Akchagylian times, many recent Mediterranean plants penetrated into the present area from the west.

Therefore, in the **Pliocene**, the piedmonts and the lower montane zone were clothed by subtropical forest vegetation. Forests of warm-temperate climate were situated higher up. Many species known from this period are still members of the modern forest flora. In the beginning of the Pliocene, within the territory of West Georgia and adjacent areas (e.g., Turkey), a Colchic refuge was formed for many relics of mesophytic forest flora. This was a direct result of the warm and humid climate of this territory. The Colchic refuge was of great importance as one of the most stable "shelters" for relic species during the Late Pliocene and especially in the Pleistocene, including the Ice Ages. Many species, which continue to exist in the Colchis died out many millions of years ago on the territory of West Eurasia. Members of the Colchic flora are such relics and/or neoendemics as *Betula medwedewii, Quercus pontica, Rhamnus imeretina, Hedera colchica, Pterocarya pterocarpa, Laurocerasus officinalis, Arctostaphylos caucasica, Rhododendron ponticum, Rh. ungernii* and *Rh. smirnowii*. At present, their relatives with a similar autecology have mainly survived in the mountains exposed to the summer monsoon in eastern and south-eastern Asia, in the Appalachians of North America, etc.; for instance, the genus *Epigaea* is represented now by just three species, one of which is distributed in Japan, the second in North America and the last in Ajara (Georgia) and Lazistan (N. Turkey).

2.2 The Quaternary

Thus, having surveyed the Tertiary history of the Georgian flora, we now come to the **Pleistocene**. During the Pleistocene, most of the thermophilic elements vanished from the flora of Georgia. The role of cold-resistant plants considerably increased, at the same time vertical shifts of the vegetational belts appeared, especially during the glacial periods.

Another event, which might have occurred during the Pleistocene is the development of boreal elements. Though the flora of Georgia was strongly influenced by the neighbouring floristic centres of Eastern Mediterranean and SW. Asia (Near East, Anterior Asia), the process of autochthonous development provided a large number of endemic species, including present-day local endemics.

In the Early Chaudian (the **Early Pleistocene**) deposits the highest taxonomical diversity and an abundance of Taxodiaceae and Cupressaceae could be observed; these families were manifested by the following genera: "*Athrotaxis*" (=extinct Taxodiacea), *Cryptomeria, Metasequoia, Sequoia, Sequoiadendron, Taxodium; Cupressus, Chamaecyparis, Libocedrus, Juniperus.*

The dominant position was occupied by *Abies nordmanniana, Tsuga diversifolia*, whereas *Abies alba, Cedrus* aff. *libani* and some others had vanished.

In the Chaudian flora, Juglandaceae were represented by several genera: *Pterocarya, Carya, Juglans*, etc. Pollen of *Zelkova* was found in the Chaudian of Georgia.

In the Chaudian, beech forests were as widely spread as the representatives of the genus *Tilia*. Chaudian pollen assemblages contained pollen of plants very closely allied to evergreen shrub *Fatsia japonica*, which is now confined to the forests of Japan, as well as that of *Symplocos* which is largely distributed in tropical areas. Obviously, the prevailing types of vegetation in the Early Chaudian were monodominant forests of *Tsuga, Abies,* representatives of Taxodiaceae and *Fagus.* Lower montane and piedmont belts were covered by the forests of *Juglans, Carpinus, Quercus, Zelkova*, and also by subtropical species of *Magnolia, Fatsia, Symplocos,* etc.

At the same time, many ferns and spore-bearing plants vanished together with the most thermophilic elements. In the Late Chaudian, almost all the plants of subtropical and temperate climate died out. The representatives of *Tsuga* and Taxodiaceae lost their leading position in forests.

The second half of the Chaudian was characterized by the predominance of *Picea-Abies* communities in almost all the mountain belts, whereas the lower parts were covered with broad-leaved forests (Tchotchieva 1965).

Middle Pleistocene. Many ferns, which can be found in the contemporary flora of Georgia, have been known since the old Euxine period. During that period, taxonomic diversity of *Abies*, *Cedrus*, *Picea* and *Tsuga* was reduced. Members of the Taxodiaceae family prevailed in forest communities. Within the territory of Georgia, only 6 out of 66 taxa, known from the Old Euxine became extinct.

In West Georgia, the old Euxinian depositions are covered by that of Uzunlar (Shatilova et al. 2011). The Uzunlarian ferns were represented by those species which continue to exist in present-day Georgia. A dominant position in forests was occupied by widely spread *Abies nordmanniana, Picea orientalis* with an admixture of *Cedrus deodara, Tsuga diversifolia* and *T. shatilova. The* Taxodiaceae family was represented by *Taxodium, Cryptomeria, Sequoia,* and *Glyptostrobus.* Hardwood forests consisted of the following species which have become members of the modern Georgian flora: *Fagus orientalis, Castanea sativa, Carpinus caucasica,* etc. It should be mentioned here that mesophytic species were substituted in East Georgia by those of *Pinus, Carpinus, Quercus,* etc.

In the **Late Pleistocene**, there were still a number of species which have since disappeared from the territory of Georgia. Mention should be made of the representatives of Taxodiaceae, of *Cedrus deodara* and *Carya aquatica.* Except for *Carya*, angiosperms were represented by species which are characteristic of modern Georgian flora.

Man introduced considerable changes into the vegetational cover of Georgia. Forest destruction, development of secondary meadows, cultivated lands are the results of human activities. At the same time, the Georgian flora was enriched both by cultivars and adventive weeds.

3 On the Altitudinal Differentiation of the Vegetation of Georgia

Considerable difference between the climates of East and West Georgia determined the diversity of their vegetal landscapes, as well as the structure of altitudinal zonation (See "Map of the Natural Vegetation of Georgia" 2003).

The absence of arid and semi-arid vegetation belts is characteristic of West Georgia. It accounts for a more simple profile of altitudinal zonation here, which is represented by five main belts: forests (0–1,900 m a.s.l.), subalpine (1,900–2,500 m), alpine (2,500–3,100 m), subnival (3,100–3,600 m), nival (above 3,600 m).

In East Georgia, the altitudinal zonation is more complicated. One can observe here six main belts: deserts, dry steppes and arid light forests (150–600 m), forests (600–1,900 m), subalpine (1,900–2,500 m), alpine (2,500–3,000 m), subnival (3,000–3,500 m), and nival (above 3,500 m). The borderline between the semi–arid belt and that of forests varies considerably due to the climatic conditions and the exposure of slopes. Within the forest and subalpine belts of South-Georgian Upland there are small areas occupied by semi-arid ecosystems with the prevalence of mountain-steppe vegetation.

Thus, almost all the altitudinal belts, characteristic of the Nemoral Zone of Northern Hemisphere, are represented in Georgia.

We would like to illustrate the differentiation of vegetation by vertical gradients in two strongly different parts of Georgia: the East and the West (the profiles are shown in Figs. 3.1 and 3.2a). In East Georgia, the vegetation of deserts, semi-deserts and hemi-xerophytic light forests is replaced by *Quercus iberica* forests with the admixture of *Carpinus orientalis* on southern slopes. At an altitude of 1,300 m, there is a narrow zone of *Carpinus-Fagus* forest, which is substituted higher by that of *Quercus macranthera*. Comparatively dry meadows can be observed at an altitude of 2,250–2,300 m.

On the northern slopes, the situation is different. Above the hemi-xerophytic vegetation belt, there is a subzone of *Carpinus-Quercus* forest, higher up substituted by beech forest which occupies large areas from 600 to 1,800 m. *Acer trautvetteri* is a common associate of the dominant *Fagus orientalis*. Still higher up, these communities are replaced by a birch forest with tall herbaceous vegetation confined to glades. Above the timberline, all slopes are covered with *Rhododendron*

G. Nakhutsrishvili, *The Vegetation of Georgia (South Caucasus)*, Geobotany Studies,
DOI 10.1007/978-3-642-29915-5_3, © Springer-Verlag Berlin Heidelberg 2013

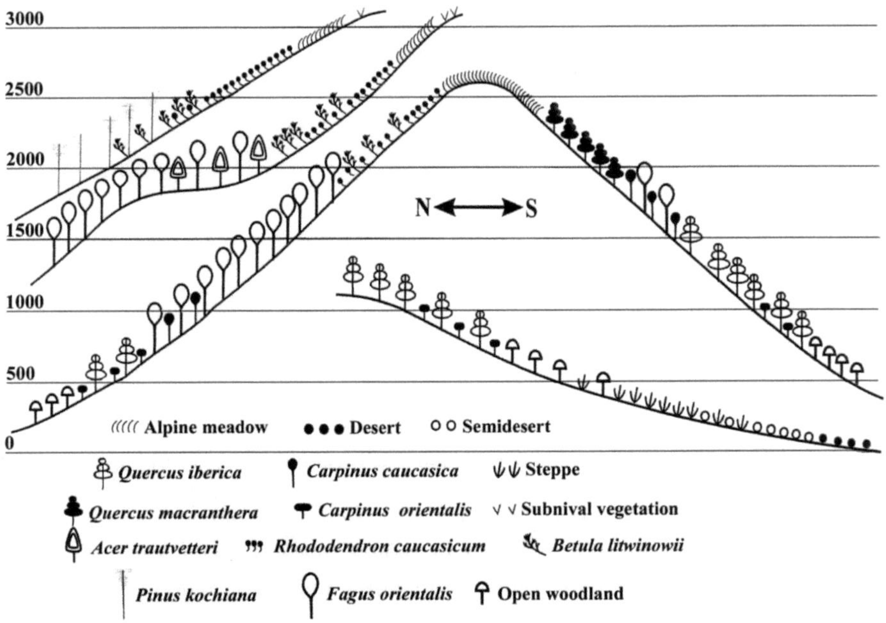

Fig. 3.1 The regularities of vertical distribution of vegetation in the East Georgia

scrub ("dekiani" is a native name of these communities, suggested for scientific use by Ketskhoveli) and carpet-like alpine meadows.

Beech forest is substituted by that of *Pinus kochiana* under more continental conditions (e.g. in Tusheti).

Unlike East Georgia, in the western part of the country, namely in South Colchis, forests begin to dominate from the sea level. The most characteristic species of these lowland forests are *Pterocarya pterocarpa, Alnus barbata, Pinus pityusa* (in north-western Colchis), etc. Southern slopes are populated by the forests of *Quercus iberica, Q. hartwissiana* and *Carpinus caucasica.* At 600–700 m, beech forests appear with admixture of *Abies nordmanniana;* spruce-fir forests (*Picea orientalis* and *Abies nordmanniana*) occur at 1,200–1,300 m. Near the timberline, *Picea orientalis* forests and that of *Betula medwediewii* are distributed. The alpine meadows occur higher up.

The oak forests are substituted by that of *Carpinus caucasica* and *Castanea sativa* on the northern slopes. Beech forests, situated higher up, form crook-stem woods at the extreme altitudinal range of their distribution.

The second variant of altitudinal zonation in West Georgia (Svaneti) is also of particular interest. *Alnus barbata* forests populate the lowland area (Fig. 3.2b). The leading species of these communities is often accompanied by *Carpinus caucasica* and *Matteuccia struthiopteris.* At 600 m, beech forests begin to dominate; their undergrowth is represented by *Rhododendron ponticum.* At about 1,300 m a.s.l., beech is usually accompanied by fir (*Abies nordmanniana*). The subalpine belt is

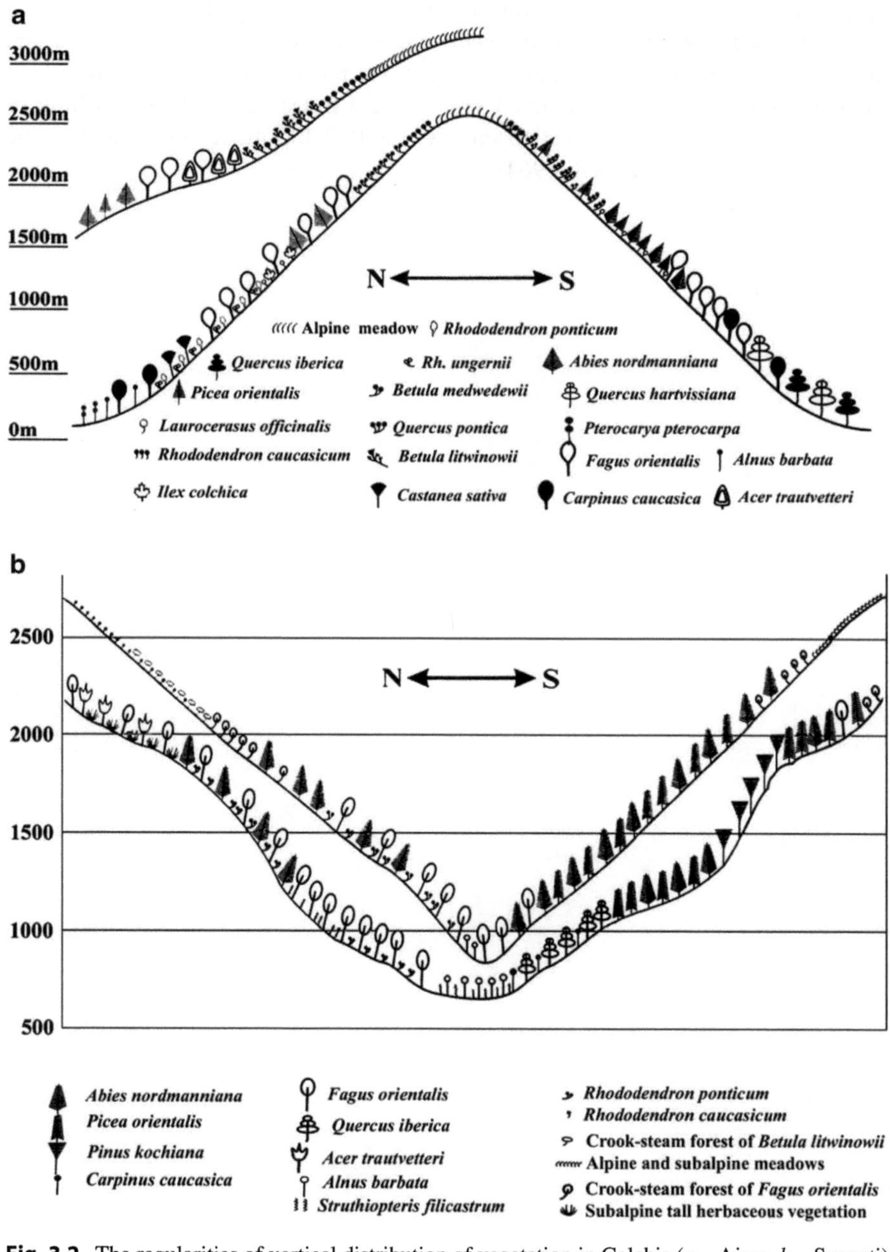

Fig. 3.2 The regularities of vertical distribution of vegetation in Colchis (*a* – Ajara, *b* – Svaneti)

characterized by the dominating position of *Acer trautvetteri* growing together with *Fagus orientalis*. The herbaceous vegetation is also characteristic of this belt. It should be noted that in several gorges the lower part of the subalpine belt is occupied by crook-stem beech forests, while the upper part is dominated by decumbent beech forests and *Rhododendron* scrub. On the southern slopes, lowland forests are substituted by those of oak, which are replaced by fir-spruce forests at about 1,000 m; rocky habitats are covered by pine forests. Near the upper limit of forests, crook-stem beech forests occur. At 1,850–2,400 m, subalpine and alpine meadows appear (Gulisashvili 1964).

On the Classification and Nomenclature of Plant Communities

4

I find it necessary to make the reader acquainted with the principles and methods of plant community classification, used to recognize syntaxa given in this book.

In the former USSR, the main approach to the classification of vegetation units was based on the dominant species. The "Formation" (higher syntaxon) was defined by the dominant species in the layer. According to the former Soviet authors, an association includes plant communities with the same set of dominants in each layer. Communities with the same dominant species in the main layer and ecological-biomorphologically similar co-dominants or dominants of secondary layers belong to groups of associations.

The above-mentioned principle of classification is known also as an ecological and physiognomical approach. The names of the syntaxa used in this book are constructed according to the below instructions.

The names of associations may be constructed in two ways. The simplest way is the following: the association is named after the names of the dominant and the subdominant species of different layers connected by a hyphen (–) (e.g., "Festuca varia-Carex meinshauseniana"). While regarding the association which consists of two dominant species in one layer, the names of such species are connected by the sign "+" (e.g., "Bromopsis variegata + Agrostis planifolia-Trifolium ambiguum").

As mentioned above, there is also a second way of construction of association names. Concerning the association where the dominant species is accompanied by the subdominant one, the name of the association should be constructed in the following way: the genus name of the dominant taxon is used with the suffix "-etum", followed by the genus name of the subdominant with the suffix "-osum" or "-etosum" (e.g. "Festucetum caricosum"). When the association consists of two dominants in the main layer, its name is constructed as follows: "Brometo-Agrostietum trifoliosum".

Names of groups of associations are constructed as follows: the genus name of the dominant taxa of the main layer has the suffix "-eta" followed by the genus name of the co-dominant or the dominant taxa of the secondary layer with the suffix "-osa" ("Fageta rhododendrosa"). In the case of formation the genus name of the dominant is used with the suffix "eta". An example is "Fageta". If we are dealing

G. Nakhutsrishvili, *The Vegetation of Georgia (South Caucasus)*, Geobotany Studies,
DOI 10.1007/978-3-642-29915-5_4, © Springer-Verlag Berlin Heidelberg 2013

with the formation dominated by two or more taxa, a compound noun is constructed using the same suffix "eta" ("Pineto-Piceeta", "Festuceta-Bothriochloeta").

Syntaxa with names constructed by the suffixes "-etum" or "-eta" mentioned in the present book, in many cases bear no syntaxonomic meaning.

The Vegetation of Semi-Deserts, Steppes and Arid Open Woodlands

5

Semi-desert and steppe vegetation, as well as arid light forests, are confined mainly to the intermountain part of East Georgia, namely to the Inner Kartli and the Lower Kartli Lowlands and the Iori plateau.

These parts of Georgia enjoy a dry subtropical climate marked by cold winter and dry hot summer. The climate of Iori plateau is more continental. The mean annual temperature varies in intermountain area of East Georgia between 9 °C and 13 °C, the mean monthly temperature in January between 1 °C and 4 °C, in August between 20 °C and 25 °C. Extreme minimum temperature falls to −20 °C to 25 °C, and extreme maximum temperature amounts to 35–40 °C. Annual precipitations range from 350 to 600 mm (Fig. 1.1).

In the area under review, the common soil types are chestnut soils, chernozems and grey–brown soils.

5.1 Vegetation of Semi-Deserts

Though typical desert vegetation is not represented in Georgia, it is remarkable that some fragments of it, namely solontchaks, can be found in the eastern part of the country (from 100 (90) to 820 m in places to 1,200–1,400 m a.s.l. in Lower Kartli Lowland, Tbilisi environs, Inner Kartli, Meskheti) (Lachashvili and Khachidze 2010). Grossheim (1948) points out that these fragments are predecessors of the semi-desert vegetation. The following species belonging to these communities should be mentioned:

Salsola ericoides	*Petrosimonia brachiata*
S. dendroides	*Nitraria schoberi*
S. nodulosa	*Atriplex cana*
Gamanthus pilosus	*Artemisia lerchiana*
Suaeda microphylla	

G. Nakhutsrishvili, *The Vegetation of Georgia (South Caucasus)*, Geobotany Studies, DOI 10.1007/978-3-642-29915-5_5, © Springer-Verlag Berlin Heidelberg 2013

Dominants of this desert vegetation are florogenetically connected with the Turan-Anterior Asian and Eastern Mediterranean floristical centres (Sakhokia 1958).

According to Lachashvili and Khachidze (2010) the following desert plant communities are present there: *Artemisia lerchiana – Salsola ericoides, Artemisia lerchiana – Salsola nodulosa, Artemisia lerchiana – Botriochloa ischaemum, Artemisia lerchiana – Reaumuria altrnifolia, Salsola nodulosa – Gamanthus pilosus – Petrosimonia brachiata, Salsola dendroides – Anabasis aphylla, Salsola dendroides – Tamarix ramosissima, Salsola dendroides – Kalidium caspicum, Gamanthus pilosus – Eremopyrum orientale – E. distans, Salsola dendroides – Artemisia lerchiana – Alhagi pseudalhagi* (Fig. 5.1a). The following ephemeral species are well represented in the above listed plant communities:

Anthemis candidissima	*Torularia torulosa*
Bombycilaena erecta	*Herniaria hirsuta*
Calendula gracilis	*Astragalus asterias*
Garhadiolus angulosus	*Vicia cinerea*
Koelpinia linearis	*Adonis aestivalis*
Arnebia decumbens	*Veronica amoena*
Alyssum desertorum	*Valerianella rimosa*
Lepidium vesicarium	*Bromus japonicus*
Leptaleum filifolium	*Rostraria glabriflora*
Sterigmostemum incanum	*Rochelia disperma*

From the ephemeroids the following are noteworthy:

Poa bulbosa var. *vivipara*
Allium rubellum
Catabrosella humilis

The following ephemeral plants grow on strongly saline soils:

Aizoon hispanicum	*Parapholis incurva*
Eremopyrum ssp.	*Spergularia diandra*
Psylliostachys spicata	

The presence of ephemers and ephemeroids are characteristic for desert vegetation, including: *Poa bulbosa* var. *vivipara, Catabrosella humilis, Bromus japonicus, Eremopyrum orientale, Alyssum desertorum, Helianthemum salicifolium,* etc. The above mentioned plants should not be regarded as xerophytes because their short duration is connected with the vernal and autumnal vegetation period. In general, these plants are mostly of mesic nature.

The above communities are characterized by the presence of species of bryophytes (*Tortulla sp., Barbula sp.*), lichens (*Toninia caeruleonigricans, Collema tenax, etc.*) and algae (species of *Nostoc*, etc.).

Fragments of desert communities dominated by *Nitraria schoberi* occur in Inner Kartli, Kiziki and Meskheti. Some variants of erosional-desert vegetation are to be

Fig. 5.1 (continued)

found in the Iori plateau (Kiziki, East Georgia) (Khintibidze 1990; Lachashvili and Khachidze 2010). Vegetation is very poor on eroded slopes with washed soil cover and is manifested by individual species of *Stipa szovitsiana, Artemisia lerchiana,* etc. The following species with flowering period in spring can be mentioned: *Tulipa eichleri, Allium rubellum,* etc.

Stipa szovitsiana, Astragalus xiphidium, Salsola nodulosa grow on the loess-like loams near the ravines and river beds, where the soil cover is absolutely denuded.

From the viewpoint of many botanists wormwood *(Artemisia lerchiana)* communities belong to desert vegetation, while others regard them as semi-desert communities. They play a considerable part in landscape formation in East Georgia, namely on the Iori plateau (Eldari) and in Lower Kartli.

Fig. 5.1 (**a**) Artemisietum lerchianae, Chachuna, East Georgia (Photo O. Abdaladze), (**b**) Stipeta (*S. pennata*) in Tbilisi vicinity, East Georgia (Photo Sh. Sikharulidze), (**c**) *Iris iberica*, East Georgia (Photo M. Akhalkatsi)

The climate of Iori plateau is dry subtropical, the dominating soil types here are grey–brown soils and chernozems.

Artemisia lerchiana is the dominant of wormwood communities. Besides the monodominant wormwood communities, Artemisieto-Salsoletum dendroides can also be found in this area.

Wormwood/salt-wort communities are restricted to the clay-brown soils and chernozems. Floristically, these communities are very poor (they include just 26 species, according to Sokhadze and Sokhadze 1991). The dominating position is occupied by annuals and ephemers (*Eremopyrum orientale, Lepidium vesicarium,* etc.); in communities with the preponderance of *Artemisia lerchiana, Stizolophus coronopifolius* is a common species.

Artemisia – Caragana grandiflora communities occur in poorly populated alluvial and grey–brown soils. According to Sokhadze and Sokhadze (1991), only 24–26 species participate in the formation of these communities. The following species, peculiar for this type of vegetation should be mentioned: *Salsola ericoides, S. dendroides,* etc. The commonest ephemers are *Lepidium vesicarium.* Examples of rare species are *Spergularia diandra, Calendula persica, Tetradiclis tenella, Aizoon hispanicum.*

Wormwood communities are widely distributed in Eldari, occupying both flat and hilly locations on salty greyish–brown soils. Floristically, these communities are also poor (about 30 species). Wormwood *(Artemisia lerchiana)* is characterized by wide ecological amplitude. The coenotypes of these communities are mainly ephemers, especially *Poa bulbosa* var. *vivipara, Bromus japonicus, Rostraria glabriflora, Medicago minima, Torularia contortuplicata.*

Bothriochloa-wormwood communities require special attention. They are the commonest vegetation type in Eldari lowland, they occur on grey–brown, skeleton and stony soils. The coverage is about 30–50 %. Wormwood communities in this area are replaced by *Bothriochloa,* due to the reduction of salt in washed out soils and increasing degression of pastures. The flora of these communities comprises over 30 species with predominance of ephemers and ephemeroids. Examples are *Poa bulbosa* var. *vivipara, Trachynia distachya, Lappula echinata, Medicago minima, Velezia rigida,* etc. (Sokhadze and Sokhadze 1991; Lachashvili and Khachidze 2010).

Pure variants of wormwood communities are widely distributed in the Caucasus, namely Georgia, but there also occur intermediate types, mixed diffusely or completely with the variants of saltwort deserts.

Wormwood communities with ephemers are found in Gardabani district (Lower Kartli). They are dominated by the following ephemers: *Adonis aestivalis, Astragalus brachyceras, Koelpinia linearis, Medicago minima, Queria hispanica,* etc.

Communities with less desert-like nature are developed under more humid conditions and on rich soils (Iaghluja, Lower Kartli). The leading species of wormwood in these communities is accompanied by the following perennials: *Salsola dendroides, Bothriochloa ischaemum, Limonium meyeri,* etc. The geophytes are represented by the species of *Iris, Gagea, Tulipa.*

Wormwood is not used for fodder, but its ephemeral synusium is important as natural grassland.

5.2 Steppe Vegetation

Steppe vegetation largely contributes to the formation of foothill landscapes (300–700 m a.s.l.) of East Georgia. It belongs to the second altitudinal belt above the desert vegetation zone.

Present-day expansion of steppes is due to the anthropogenic influence on forests, arid light forests and even on secondary shrubwoods (Sakhokia 1958).

The dominant species of steppe vegetation – *Bothriochloa ischaemum* – is a representative of the subtropical-tropical genus *Bothriochloa* widely distributed in both hemispheres. Many Georgian botanists are inclined to attribute *Bothriochloa* to semi-steppes. Typologically, they are very diverse.

The following plant communities of Bothriochloeta are distinguished by Lachashvili and Khachidze (2009):

Botriochloa ischaemum-Glycyrrhiza glabra
Botriochloa ischaemum-Festuca valesiaca
Botriochloa ischaemum-Stipa lessingiana
Botriochloa ischaemum-Paliurus spina-christi
Botriochloa ischaemum-Artemisia lerchiana
Botriochloa ischaemum-Onobrychis kachetica
Botriochloa ischaemum-with ephemerals

Some of them are described below.

Bothriochloa communities mainly occur on chestnut soils and are rare on chernozems.

Glycyrrhizieto-Bothriochloeta. These communities are mostly distributed in areas of 500–600 m a.s.l. The above communities are rather rich floristically, e.g. in an area of 100 m^2 the average number of species is 65 (Sokhadze and Sokhadze 1991). Besides the edificator and sub-edificator, the following species should be mentioned: *Koeleria cristata, Phleum phleoides, Medicago caerulea, Bromus japonicus, Trifolium campestre, Vicia angustifolia, Crucianella angustifolia, Arabidopsis thaliana, Androsace elongata.*

Bothriochloa communities with xerophytes are restricted to slopes of eastern and western exposures; they occur mainly on thin chestnut soils (600–700 m a.s.l.). The following herbs play an essential part in these communities: *Teucrium nuchense, Galium verum, Picris strigosa, Scorzonera biebersteinii, Inula britannica, Aster ibericus, Stachys atherocalyx*. Among *Fabaceae* species a mention should be made of *Onobrychis kachetica, O. cyri, Astragalus brachycarpus*. An example of the commonest species is *Festuca sulcata*.

The early vernal vegetation is dominated by the following ephemers and ephemeroids: *Arabidopsis thaliana, Anthemis candidissima, Crocus adamii, Merendera trigyna, Muscari caucasicum, Gagea commutata*. These communities include endemics of Georgia and Caucasus, such as: *Polygala transcaucasica, Aster ibericus, Cephalaria media, Thymus tiflisiensis*, etc. The following species are to be found in Glycyrrhizieto-Bothriochloetum varioherbosum:

Bothriochloa ischaemum	*Bromus japonicus*
Dactylis glomerata	*Festuca sulcata*
Koeleria gracilis	*Melica transsilvanica*
Phleum phleoides	*Stipa capillata*
Trisetum pratense	*Carex bordzilowskii*
Glycyrrhiza glabra	*Medicago caerulea*
Trifolium arvense	*Achillea nobilis*
T. campestre	*Allium pseudoflavum*
Vicia angustifollia	*Crepis marschallii*
V. hirsuta	*Bellevalia wilhelmsii*

(continued)

V. tetrasperma	*Crucianella angustifolia*
Daucus carota	*Dianthus inamoenus*
Echinops sphaerocephalus	*Eryngium campestre*
Falcaria vulgaris	*Galium verum*
Gladiolus italicus	*Hieracium pilosella*
Linum corymbulosum	*Muscari caucasicum*
Ornithogalum ponticum	*Picris strigosa*
Polygala transcaucasica	*Potentilla recta*
Tragopogon graminifolium	*Rumex tuberosa*
T. tuberosus	*Turritis (Arabis) glabra*
Phlomis pungens	*Xeranthemum squarrosum*
Ph. tuberosa	*Ranunculus illyricus*

Bothriochloeta ephemerosa. These communities are confined to stony southern slopes of chestnut soils. The following ephemers can be mentioned: *Alyssum campestre, Callipeltis cucullaria, Sideritis montana, Meniocus linifolius, Ziziphora capitata, Trigonella spicata*. Examples of geophytes are *Juno caucasica, Allium atroviolaceum,* etc.

The less xeric variants of *Bothriochloa* communities are Bothriochloeta pratoherbosa, which populate depressions and dells with chernozem-like soils. Besides the leading species of *Bothriochloa*, the following plants can be found: *Agropyron cristatum, A. pectinatum, Elytrigia (Agropyron) repens* var. *glaucescens*.

Festuceta-Bothriochloeta. These communities are restricted to the slopes of hills. Associate of *Bothriochloa* is *Festuca sulcata*. Fragments of these communities have remained only in Lower Kartli (Iaghluja).

Stipa-Bothriochloa coenoses are also found in the above-mentioned area; they are restricted to the northern mountain slopes, ridges, stony habitats. Most of the soils in these areas are grey–brown, thin, skeleton and saltless.

The leading species of these communities are accompanied by shrubs like *Paliurus spina-christi, Rhamnus pallasii,* etc. Herbs are represented by steppe elements. Examples are *Stipa* spp., *Onobrychis* spp., *Phleum phleoides*. The number of ephemers is relatively small.

Alyssum campestre
A. desertorum
Androsace elongata
Arabidopsis thaliana
Callipeltis cucularia
Echinaria capitata
Helianthemum lasiocarpum
H. salicifolium
Kohlrauschia prolifera
Linum corymbulosum
L. nodiflorum

(continued)

Pterotheca sancta
Salvia viridis
Scabiosa micrantha
Trigonella spicata
Vicia angustifolia
Ziziphora capitata

The following list of species mentions members of Bothriochloeta varioherbosum:

Stipa lessingiana	*Tragopogon pusillus*	*Thalictrum minus*
S. capillata	*Campanula hohenackeri*	*Filipendula vulgaris*
S. pulcherrima	*Crucianella angustifolia*	*Potentilla recta*
Phleum phleoides	*Trigonella spicata*	*Veronica multifida*
Medicago coerulea	*Bromus japonicus*	*Bilacunaria caspia*
Inula germanica	*Bothriochloa ischaemum*	*Seseli grandivittatum*
Galium verum	*Festuca sulcata*	*Cephalaria media*
Leontodon asperrimum	*Koeleria macrantha*	*Xeranthemum squarrosum*
Thymus tiflisiensis	*Cleistogenes bulgarica*	*Linum corymbulosum*
Picris strigosa	*Astragalus brachycarpus*	*L. nodiflorum*
Pimpinella aromatica	*Stachys iberica*	

According to Sakhokia (1958), *Bothriochloa* communities and their edificators are closely related to savannas florogenetically, rhythmologically and by the character of soding; though it is not the case, when the above communities get compared with the typical *Stipa* steppes. This opinion is confirmed by the fact that *Bothriochloa* communities are combined with arid light forests, which, in the past, had zonal significance in the Caucasus. The same author has proposed the term "savannoid vegetation" for the complex of *Bothriochloa* communities and arid light forests.

Typical steppe vegetation is illustrated by the comunities of *Stipa* (*S. tirsa, S. pennata, S. lessingiana, S. pulcherrima, S. capillata*) (Fig. 5.1b). Participation of the above communities in landscape formation is insignificant. *Stipa* communities are of secondary origin. They favour the areas which have previously been occupied by mesic mountain forests, shrubwoods and herbaceous vegetation. Very often in the ridges of mountain ranges, *Bothriochloa* steppe becomes replaced by the communities of *Stipa*, including more xeric species, such as *S. pulcherrima, S. lessingiana, S. pontica,* and *S. capillata.* These communities are characterized by the presence of the following xeric species: *Seseli grandivittatum, Teucrium nuchense, T. polium, Thymus tiflisiensis, Scorzonera eriosperma, Psephellus carthalinicus, Carex bordzilowskii, Tulipa biebersteinii, Tulipa eichleri, Iris iberica* (Fig. 5.1c).

Stipa tirsa and its communities are confined to thick chernozems with a higher amount of moisture in soil. *S. tirsa* is known as a soding plant. Communities of *S. tirsa* are to be found in Gareji steppe. The associates of the dominant *Stipa* are *Glycyrrhiza glabra, Medicago coerulea, Koeleria cristata,* etc.

Drier habitats are occupied by *Stipa joannis* and *S. lessingiana*. The plant communities, which the above-mentioned species enter as the leading ones, do not include *Stipa tirsa*. Chernozems and grey–brown soils, mostly calcareous, are typical for this area. Slightly destroyed communities, restricted to the northern slopes, are characterized by the presence of *Dianthus subulosus*, and *Pyrethrum corymbosum*; among other associates is *Bothriochloa ischaemum*. Under the influence of intense animal grazing *Bothriochloa* becomes a co-edificator of the species of *Stipa*, whereas the community gets floristically poorer.

In rocky areas, such as the vicinity of David Garedja Monastery, the communities are dominated by the species of *Bothriochloa, Festuca*, etc., while *Stipa lessingiana* is not found here (Sokhadze and Sokhadze 1991).

Festuceto valesiaca-Botriochloeta is one of the widespread plant communities with the following set of perennials:

Teucrium polium	*Scorzonera biebersteinii*
Thymus coriifolius	*Stachys atherocalyx*
Achillea biebersteinii	Of the annuals the following are noteworthy:
Eryngium campestre	*Alyssum alyssoides*
Euphorbia seguieriana	*Helianthemum lasiocarpum*
Falcaria vulgaris	*H. salicifolium*
Onobrychis cyri	*Kohlrauschia prolifera*
Koeleria cristata	*Scabiosa micrantha*
Medicago coerulea	*Trifolium campestre*
Poterium polygamum	

Forb steppes (with *Stipa tirsa*) are rather rich in species; besides the representatives of *Stipa*, various forbs participate in the composition of this community.

It should be pointed out, that mountain steppes are peculiar only for South Georgia. Their altitudinal distribution ranges from 1,800 to 2,500 m a.s.l. They occur on leached chernozems. Mountain steppes are represented by that of *Festuca-Stipa* and meadow steppes.

Festuceto (*F. sulcata*)-Stipeta (*S. capillata*). These are developed on flat grounds and southern slopes. They are characterized by high presence of *Dactylis glomerata, Stipa tirsa, Phleum phleoides, Trifolium alpestre, Medicago dzhavakhetica*, etc. Geophytes are exemplified by species of *Gagea, Muscari*, etc.

Meadow steppes are situated on the northern slopes. *Stipa tirsa* is an edificator of these communities. Tall dense herbage with a large amount of species is typical for this variant of steppes; a mention should be made of the high-presence of dicots (*Betonica macrantha, Aster ibericus*, etc.). Besides, geophytes are well developed here.

Difference of opinion has arisen upon the issue of a possible connexion between the northern plain steppes (South Russia) and those of the Transcaucasian mountains.

Grossheim (1948) indicates that South Russian and Transcaucasian mountain steppes, being separated from each other nowadays, have formed an integral steppe massif in the past. The formation of the above massif was connected to the process of peneplainization of the Caucasus during more or less xerothermic periods. The uplifting of the Greater Caucasus caused separation of the steppe massif from its northern part; it has been preserved under more or less favourable conditions in mountainous areas. Probably, the above mentioned events account for the great floristic and phytocoenotic similarity between the steppes of South Russia and those of the Transcaucasian highlands.

5.3 Arid Open Woodlands

Xeric (arid) open woodlands favour foothills and plains under the dry climatic conditions of East Georgia among steppe and desert vegetation. These communities are provided by the xerophytic woody plants on the background of drought-resistant grass cover (Fig. 5.2a–c).

The arboreal components of open woodlands or "light forests" (a term used by some Caucasian botanists) never make dense stands. In Georgia, these forests are well developed on the territory between the Alazani and the Iori river valleys, near the place Vashlovani, which holds an area of 5,000 ha.

The following types of communities of open woodland can be distinguished: pistache-woodlands, juniper open woodlands, communities dominated by species of *Pyrus* and *Celtis*.

***Pistacia mutica* communities** occur on cinnamon-coloured and chernozem-like soils. These plant communities belong to West-Iranian type. The undergrowth of these open woodlands is represented by the following shrubs: *Paliurus spina-christi, Rhamnus pallasii, Cotinus coggygria, Cerasus incana, Lonicera iberica, Pyrus salicifolia Colutea orientalis, Caragana grandiflora, Ephedra procera, Punica granatum, Rhus coriaria*. The ground layer is made of xerophytic species; it is rather rich in species (*Bothriochloa ischaemum, Festuca sulcata,* species of *Stipa*, etc.). One can distinguish within the pistache-woodland the communities with: *Bothriochloa, Stipa*.

It is also possible to find pure *Pistacia*-woodlands and the pistache-woodlands with the admixture of single trees and shrubs (*Ulmus minor, Celtis caucasica, Pyrus salicifolia,* etc.).

Open woodland-communities of *Pyrus salicifolia, P. georgica,* etc. are formed by the xeric species of pear, which join the sections *Xeropyrenia* and *Argyromalon.*

From the coenotic point of view, the commonest types of the above-mentioned woodlands are the communities with shrubs and steppe elements (Ketskhoveli 1960).

Juniperus open woodlands occur on, steep northern slopes of low ranges of southeast part of Iori plateau and the environs of Mtskheta. In both places, the dominants of forests are *Juniperus foetidissima* (East-Mediterranean species) and *J. polycarpos* (species of Asia Anterior–South-West Asia). In southeast part of Iori plateau they are accompanied by *Pistacia mutica.*

Fig. 5.2 (continued)

Besides the above mentioned species, the East-Mediterranean *J. rufescens* and the Caucasian *J. oblonga* occur in *Juniperus* open woodlands as undergrowth.

Open woodlands occur on grey–cinnamon soils. The climate in the area of their distribution varies between dry subtropical and warm temperate with not very hot summer season. The amount of annual precipitation is 400 mm in Vashlovani. In the environs of Mtskheta, the climate is warmer temperate with hot summer and the amount of precipitation equals to 600–700 mm p. a. (Sakhokia 1958; Lachashvili et al. 2004).

Open woodlands of *Celtis caucasica* hold rather small areas. Usually, solitary plants or groups of plants of *Celtis caucasica* can be found.

Natural vegetation in the east is preserved in the Vashlovani Reserve in Kiziki, at the far southeastern end of Georgia, near the Azerbaijani border (Lachashvili et al. 2004).

c

Fig. 5.2 (**a**) Arid open woodland (Junipereta), Vashlovani National Park, East Georgia (Photo O. Abdaladze), (**b**) Arid open woodland (Pistacietum), Vashlovani National Park, East Georgia (Photo O. Abdaladze), (**c**) *Pistacia mutica*, Vashlovani National Park, East Georgia (Photo O. Abdaladze)

The most interesting steppe type is meadow-steppe, the relatively tall (to 1 m), forb-rich grassland type of the Ukraine (Walter 1974; Walter and Breckle 1986) equivalent to the true (i.e. tall-grass) prairie of North America. A sample from a small but representative patch of meadow-steppe in the Vashlovani Reserve is shown by the relevé in Table 5.1. *Botriochloa* (=*Andropogon*) *ischaemum* is the dominant in this stand, as in most of the Georgian steppes. The relevé also, however, includes *Stipa pulcherrima*, one of several "feather grass" species characteristic of Ukrainian to Middle Asian steppes, as well as some *Paliurus spina-christi* shrubs, as characteristic of "savannoid" steppe areas extending westward at least as far as Tbilisi. Meadow-steppe undergoes a continuing metamorphosis during the growing season, mainly from April to June, as early forbs are replaced by taller forbs and grasses, which finally yield to the dominants of the early-summer aspect (Walter 1968; Walter and Breckle 1986). As with other meadow-steppe areas, grasses make up most of the biomass. Only 12 of the 63 species in the relevé, however, are grasses.

The other important landscape type from the east is the *Pistacia-Juniperus* "arid woodland" (Fig. 5.2a–c), an open woodland on foothills and plains in the driest parts of eastern Georgia. Areas of arid woodland are also well represented in the Vashlovani Reserve, a sample of which is shown in Table 5.2. The main structural elements are three *Juniperus* species plus deciduous *Pistacia mutica*, all of which grow in the form of small trees. Spiny *Paliurus spina-christi* is a major understorey shrub, as are *Jasminum fruticans*, *Berberis iberica*, *Rhamnus pallasii*, *Cerasus* spp., and younger individuals of the juniper species. On flatter terrain these woodlands generally constitute parklands of wooded patches (20–50 % of the area) embedded within a grassy Stipetum matrix. This grassy matrix is not completely represented

Table 5.1 Meadow-Steppe in Easternmost Georgia (relevé G-16) (Box et al. 2000) (Location: Vashlovani Nature Reserve, border with Pantishara Gorge (31 May 1999))

S	2.0 m	5 %	550 m, aspect: 5–10° to N	
H	0.8 m	85 %	10 × 30 m	
S	1.1	*Paliurus spina-christi*	1.1	*Cotinus coggygria*
	+.2	*Asparagus officinalis*	+	*Convolvulus cantabrica*
	+	*Lonicera iberica*		
H	4.4	*Botriochloa ischaemum*	2.2	*Koeleria cristata*
	2.2	*Stipa pulcherrima*	1.1	*Stipa capillata*
	1.1	*Cleistogenes bulgarica*	1.1	*Festuca valesiaca*
	1.1	*Bromus japonicus*	1.1	*Phleum phleoides*
	+	*Hordeum crinitum*	+	*Aegilops triuncialis*
	+	*Trachynia distachya*	(+.2)	*Dactylis glomerata*
Legumes				
	2.2	*Medicago coerulea*	2.2	*Onobrychis kachetica*
	1.1	*Medicago minima*	+	*Astragalus brachycarpus*
	+	*Onobrychis cyri*		
Forbs				
	2.2	*Thymus tiflisiensis*	2.2	*Galium verum*
	1.2	*Filipendula hexapetala*	1.1	*Poterium (=Sanguisorba) polygamum*
	1.1	*Achillea nobilis*	1.1	*Potentilla recta*
	1.1	*Scorzonera biebersteinii*	1.1	*Tragopogon tuberosus*
	1.1	*Jurinea blanda*	1.1	*Plantago lanceolata*
	1.1	*Helianthemum salicifolium*	1.1	*Hypericum perforatum*
	1.1	*Teucrium nuchense*	1.1	*Euphorbia seguieriana*
	1.1	*Teucrium polium*	1.1	*Inula germanica*
	1.1	*Scutellaria orientalis*	1.1	*Polygala transcaucasica*
	1.1	*Onosma armeniaca*	+	*Crepis marschalii*
	+	*Hippomarathrum crispum*	+	*Veronica multifida*
	+	*Orobanche* cf. *speciosa*	+	*Seseli grandivittatum*
	+	*Picris strigosa*	+	*Dianthus inamoenus*
	+	*Bellevalia wilhelmsii*	+	*Malabaila dasyantha*
	+	*Sisymbrium loeselii*	+	*Linum austriacum*
	+	*Echinops sphaerocephalus*	+	*Muscari caucasicum*
	+	*Crinitaria villosa*	+	*Ziziphora serpyllacea*
	+	*Thalictrum minus*	+	*Reseda lutea*
	+	*Cuscuta speciosa*	+	*Phlomis pungens*
	+	*Eryngium campesetre*	+	*Falcaria vulgaris*

Total number of species: 63

in the relevé (Table 5.2) but was similar to the meadow-steppe site of Table 5.1, dominated, as elsewhere in the region, by *Botriochloa ischaemum*, along with *Stipa* species such as *S. pulcherrima*. Although arid woodland may be considered a west Iranian community, dry woodlands and related shrublands are important throughout much of especially eastern Georgia, extending westward to Tbilisi and on through

Table 5.2 Arid open Woodland Mosaic in easternmost Georgia (relevé G-17) (Box et al. 2000) (Location: Vashlovani Nature Reserve, down road from field station (31 May 1999))

T	8 m	20 %	550 m, slope: 5–20°, rollintg	
S	3 m	30 %		
H	0.5 m	50 %	50 × 50 m	
T	2.4	*Pistacia mutica*	2.4	*Juniperus foetidissima*
	1.1	*Juniperus polycarpos*	1.1	*Juniperus rufescens*
S	3.4	*Paliurus spina-christi*	3.4	*Jasminum fruticans*
	2.4	*Juniperus foetidissima*	1.1	*Juniperus polycarpos*
	1.1	*Juniperus rufescens*	1.1	*Rhamnus pallasii*
	1.1	*Berberis iberica*	1.1	*Cerasus incana*
	1.1	*Cerasus microcarpa*	1.1	*Lonicera iberica*
H	3.4	*Achnatherum bromoides*	1.1	*Achillea nobilis*
	1.1	*Polygala transcaucasica*	1.1	*Dactylis glomerata*
	1.1	*Potentilla recta*	1.1	*Cleistogenes bulgarica*
	1.1	*Falcaria vulgaris*	1.1	*Dictamnus caucasicus*
	1.1	*Campanula hohenackeri*	1.1	*Rumex tuberosus*
	1.1	*Koeleria cristata*	+	*Helianthemum salicifolium*
	+		+	*Silene boissieri*

Total number of species: 25

central Georgia along the Mtkvari river valley to the Meskheti region near the Turkish border.

5.4 Hemixerophytic Shrubwoods

Interzonal drought-resistant shrubwoods occur almost in every mountain belt, except in the highlands. In the geobotanical literature, different terms are applied to this type of vegetation, for example, "thorny shrubwoods", "shibliak", etc.

Many authors consider the above communities to be analogues of the Mediterranean shibliak. This affinity confirmed by Rikli (1943) – an eminent specialist of the Mediterranean vegetation.

About 25–30 species contribute to the formation of shibliak. Mediterranean shibliak consists of the following species: *Paliurus spina-christi, Berberis vulgaris, Cotinus coggygria, Punica granatum, Carpinus orientalis.*

Communities dominated by the Christ's thorn are the commonest ones in the area under review; a mention should be made of Paliureto Bothriochloeta, Spiraeeto (*Spiraea hypericifolia* (Fig. 5.3))-Paliureta, as well as hemixerophytic mixed-shrub shibliak with *Paliurus spina-christi, Crataegus orientalis, Lonicera iberica,* etc. (Figs. 5.4 and 5.5a).

On dry slopes endemic *Amygdalus georgica* can rarely be found (Fig. 5.5b).

The most xeric variant of shibliak is confined to mother rocks and stony slopes (*Rhamnus pallasii, Caragana grandiflora, Atraphaxis spinosa, Ephedra procera*).

Fig. 5.3 *Spiraea hypericifolia*, East Georgia (Photo N. Lachashvili)

Fig. 5.4 Bothriochloeto-Paliuretum, Tbilisi surroundings (Photo O. Abdaladze)

Within this certain area, the following types of shibliak are distinguished: (1) primary shibliak, restricted to slopes and steppe dell belts, as well as to beams, ancient detrital cones and naked rock outcrops; (2) the remnants of open woodland dominated by *Pyrus* and *Pistacia*; (3) secondary shibliak, the origin of which is connected with the destruction of forests (Sakhokia 1958).

In East Georgia, phrygana, the second type of hemixerophytic Mediterranean vegetation, is distributed. In order to outline the difference between the Caucasian phrygana and the true Mediterranean one, many botanists define it by the term

Fig. 5.5 (**a**) *Celtis caucasica* (Photo O. Abdaladze), (**b**) *Amygdalus georgica*, Tbilisi surroundings (Photo O. Abdaladze)

"phryganoid vegetation". In the classical sense, phrygana is regarded as a community dominated by dwarf xeromorphic shrubs and subshrubs with an admixture of therophytes and geophytes. In the Caucasus and particularly in Georgia, it is often found among the xeric mountain vegetation.

Among its coenological formations, the following should be noted:

1. Tragacanthic communities, dominated by thorny *Astragalus* species *(A. denudatus, A. microcephalus,* etc.*)* and species of *Acantholimon (A. lepturoides,* etc.);
2. Tomillares, low shrub and semi-shrub formations with the dominance of *Thymus tiflisiensis* (and other species of *Thymus*), *Teucrium spp., Salvia garedji,* etc.

6 Forest Vegetation of Georgia

Forest is the prevailing type of vegetation in Georgia. The forest area makes up 36.7 % of the country's total land area.

During the Holocene, when the climatic conditions were similar to those of nowadays, the entire territory of Georgia was occupied by forests. Very surprisingly, today beech *(Fagus orientalis)* occupies a dominating position, namely 51 % of the total forest area. The following species are further dominants of Georgian forests: *Abies nordmanniana* (8.5 % of the forest-coverage). *Quercus iberica* and other representatives of *Quercus* (10.5 %), *Picea orientalis* (6.3 %), *Pinus kochiana* (3.6 %), *Alnus barbata* (3 %), *Castanea sativa* (2.1 %), *Betula litwinowi* and other *Betula* species (*c.* 2 %). Negligible areas are covered by *Carpinus caucasica, Tilia begoniifolia, Acer platanoides, A. trautvetteri, Fraxinus excelsior,* etc. (Dolukhanov 2010).

As it was pointed out, the plains and mountain slopes in West Georgia are covered by forests from the very sea level, while lowland woods in the eastern part of the country are peculiar for riversides. At the same time, the plateau of Javakheti Upland (South Georgia) is absolutely devoid of forests.

Located in the eastern (E, SE, NE) portion of the Black Sea catchment basin the climate of the Colchis region is characterized as having moderately warm (24–25 °C summers and cool 4–6 °C winters) with abundant annual precipitation (1,800–2,200 mm up to 4,500 mm).

Colchic forests like Hyrcan forests (Azerbaijan, Iran) are the most important relicts of the Arcto-Tertiary forests in western Eurasia with many relict and endemic plants and rare fauna. Many plants have ancient boreal affinities from the Tertiary period, and, therefore, the Caucasus is considered a global "hot spot" – an area where numerous species are highly concentrated – as recognized by Conservation International (also see www.nationalgeographic.com/wildworld/profiles/terrestrial/pa/pa0422.html) and a globally unique ecoregion as recognized by the World Wildlife Fund.

Although the region has temperature levels that exceed thresholds used in the global temperate rainforest model, and some authors consider the Colchic rainforest subtropical (Rikli 1943) but with temperate tree composition (Lavrenko 1958;

G. Nakhutsrishvili, *The Vegetation of Georgia (South Caucasus)*, Geobotany Studies, DOI 10.1007/978-3-642-29915-5_6,

Fig. 6.1 (continued)

Dolukhanov 1980), this region should qualify as temperate rainforest. For instance, Colchis cannot be considered subtropical climatically or structurally in terms of plant communities: air temperatures are lower, seasonal distribution of rainfall is largely continuous, and there is no broad-leaved evergreen forest as in subtropical regions. Further, in the explanatory text to the Map of Natural Vegetation of Europe (Bohn et al. 2003), the Colchic forests are considered humid- and warm-requiring (hygro-thermophilous) broad-leaved forests (Dolukhanov 1980; Nakhutsrishvili 1999; Doluchanov and Nakhutsrishvili 2003).

Evidence for temperate rainforest in the Caucasus is generally related to the mountainous chains located along coastlines trap a large portion of the moisture arising from oceanic air masses on their windward side. In the Caucasus, these barriers are formed by a topographical triangle created by the intersection of the

Fig. 6.1 (**a**) *Rhododendron ponticum*, (**b**) *Rhododendron ungernii*, (**c**) (1–2) *Rhododendron smirnowii*, South Colchis (Photos Z. Manvelidze)

western part of the Greater Caucasus Mountain Range (Georgia, Russia), western part of the Lesser Caucasus Mountains (Turkey, Georgia) and Likhi Ridge (bridge ridge between Greater and Lesser Caucasus, Georgia) at the Black Sea. The warm and humid climate of this region has been present since the late Tertiary, the primary reason the Caucasus has acted as a shelter for hygro-thermophilous relicts during the previous ice age. Consequently, Colchic forests along with the Hyrcanian forests are the oldest forests in Western Eurasia in terms of their origin and evolutionary history, most diverse in terms of relict and endemic woody species and tree diversity, and most natural in terms of transformation of historic structure. Both Colchis and Hyrcan, whose formation is attributed to the Upper Pliocene (Kolakovsky 1961), have a number of common features.

The total area of all forests of the Colchis region, estimated by the GIS unit of WWF Caucasus using Google images, is ~ three million hectares. There are a number of forest types belonging to this region: lowland hardwood forests; foggy gorges and mixed broadleaf forests; sweet chestnut forest (*Castanea sativa*); beach (*Fagus orientalis*) forest; dark coniferous forest; and oak woodland. But the main distinguishing feature of these forests is the half-prostrate shrubs characterized by vegetative reproduction, forming dense and high understory (up to 4 m). Forests are marked by broad-leaved evergreens, including several rhododendrons (*Rhododendron ponticum* (Fig. 6.1a), *R. ungernii* (Fig. 6.1b), *R. smirnowii* (Fig. 6.1c), the last two are local endemics of Southern Colchis), Cherry-laurel (*Laurocerasus officinalis*), Black sea holly (*Ilex colchica*), as well as deciduous mountain cranberry (*Vaccinium arctostaphylos*), and oriental viburnum (*Viburnum orientale*) (Zazanashvili 2009; Nakhutsrishvili et al. 2011).

6.1 Lowland Hardwood Forests

In the Caucasus, including Georgia, the lowland forests are spread on swamps, flood plains and in the lowlands, where local irrigation conditions or groundwaters stimulate the development of forest vegetation (Grossheim 1948).

Swamp forests occupy the Colchic Lowland with its mainly damp and poorly drained soils. The leading species of this community is *Alnus barbata*. The above forests are floristically rather rich – about 160 species can be found. The number of exclusive species is negligible; among them *Matteuccia struthiopteris*. *Alnus* forests with *Buxus colchica* are rather rare; near the upper range of distribution, *Alnus* forest with tall herbs (*Telekia speciosa, Heracleum ponticum*) is developed. Due to the extreme swampy conditions, alder becomes dwarfed and provides large hillocks, on which the following plants are developed: *Smilax excelsa* (climbing on alder), *Iris pseudacorus, Leucojum aestivum, Oenanthe abchasica*, etc. (Kolakovsky 1961).

Riparian or flood forests are to be met on river banks in both woody and woodless regions. In riparian forests of East Georgia, as well as in Colchis, *Pterocarya pterocarpa* is a typical species. Besides, these communities are characterized by presence of *Quercus pedunculiflora* (= *Q. longipes*), *Populus alba, Elaeagnus angustifolia, Tamarix ramosissima, T. hohenackeri, Hippophae rhamnoides*, etc.

Among the lianas climbing on these trees are the following: *Vitis sylvestris, Periploca graeca, Cynanchum acutum, Solanum persicum*, etc.

Lowland forests are characteristic of Kakheti, the East Georgia district, where the annual precipitation amounts to 900 mm p. a. These forests are dominated by *Quercus pedunculiflora* with the admixture of *Acer velutinum, Tilia caucasica, Fraxinus excelsior, Pyrus caucasica*. Frequently occurring as underwood in these communities are: *Crataegus pentagyna, Mespilus germanica*, etc. The following lianas are also present here: *Hedera pastuchowii, H. helix, Vitis sylvestris, Clematis vitalba, Smilax excelsa*. According to Grossheim (1936, 1948), these forests reflect the vegetation of the Hyrcanian territory which is the refuge area of Tertiary flora in the Caucasus.

6.2 Lowland Coniferous Forests

These communities are demonstrated by *Pinus pityusa* forests.

The area of distribution of *Pinus pityusa* occupies almost the whole eastern coast of the Black Sea from Mussera (southwards from Bichvinta/Pitsunda cape) to Anapa (Ukraine). This relic species is closely related to the Mediterranean *P. halepensis*, to the Crimean *P. stankevichi* and to the east-Transcaucasian *P. eldarica*. In Bichvinta it forms high-trunk pine forests, where almost all the trees are of the same age; this community holds an area of 200 hectares and is considered to be of secondary origin, due to the elevation of the sea coast and the exposure of sandy-pebble deposites of Bichvinta cape. Undergrowth of this forest is illustrated by *Cistus creticus, Ruscus ponticus, Rhododendron luteum* and *Mespilus germanica*. The following types of these communities are to be distinguished: *Pinus pityusa* forest with *Carpinus orientalis*, mixed pine-broad-leaved forest, Pitsunda pine forests of coastal area, etc. (Tumajaniv 1980).

A mention should be made of *Pinus eldarica*, which is confined to the Eilar-Ougi mountain range near the Georgian/Azerbaijan frontier.

The distribution range of Eldari pine includes the Mtkvari-Araxian desert and mountain-steppe province of the Afro-Asian desert region. The floristic complex of Eldari pine communities is greatly influenced by paleogeographical peculiarities of the present-day distribution area of *Pinus eldarica*. The associates of Eldari pine are the plants of arid open woodlands.

6.3 Mountain Forests

In the mountains of Georgia, within the forest belt, three sub-zones can be outlined: low mountain, mid-mountain and high-mountain (from 500 to 1,900 m a.s.l.). Polydominance is one of the most striking features of mountain forests (except beech woods). Timberline today varies in altitude between 2,200 and 2,750 m, but almost everywhere it has been lowered by man. In the western part of the Greater Caucasus the upper limit of woody vegetation is decreased to 350–400 m, whereas in central and eastern parts – to 500–600 m (Fig. 6.2). The same situation is in the Lesser Caucasus. Due to the impact of anthropogenic factors, the forest vegetation of the Kazbegi region (eastern part of the Central Caucasus) has almost completely been destroyed.

Below, we try to characterize the main formations of forest vegetation of Georgia.

6.3.1 Beech Woods

Beech woods of the Caucasus and particularly Georgia, are dominated by the ancient east Mediterranean species of the Oriental beech *Fagus orientalis* (Fig. 6.3a, b). Its distribution range comprises Mountainous Crimea, the Greater and Lesser Caucasus, Pontian Mountains and the Caspian sector of Iran.

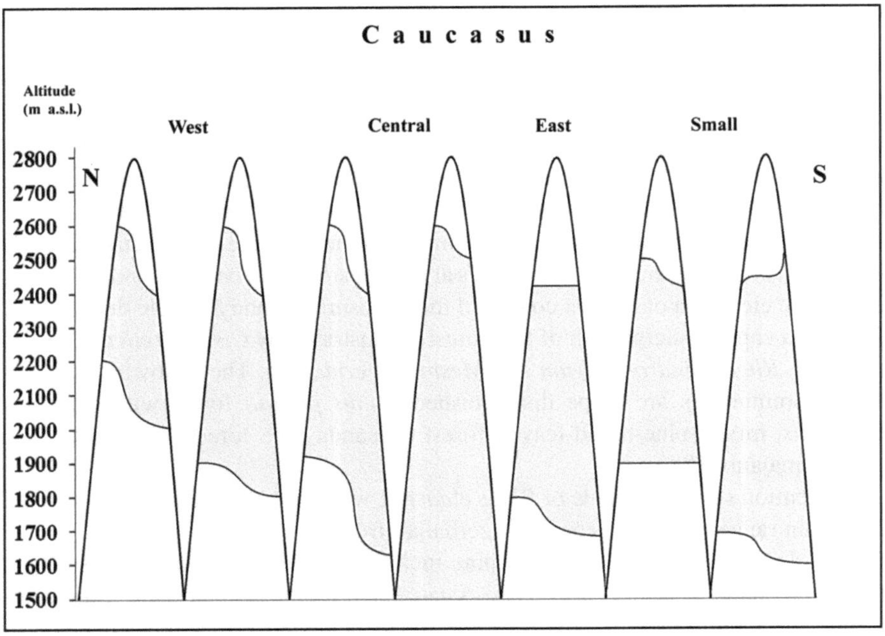

Fig. 6.2 The upper limits of forests

The lower limit of beech forests ranges from sea level (Colchis) to 1,000 m a.s.l., sometimes rising to 1,200–1,300 m (under the most arid conditions). The upper limit varies in altitude between 1,400 and 2,000–2,200 m, and this is the most constant boundary of their altitudinal distribution, which usually depends on the humidity of climate. This accounts for the fluctuations of the upper limit of beech woods in different districts of Georgia. For instance, in moist Colchis, the scope of distribution varies from sea level to 2,200–2,380 m, whereas in the eastern part of the country (Alazani basin) the lower limit of beech forests comes to approximately 350–400 m. In Trialeti mountains (southern Georgia) beech woods appear at an altitude of 800–900 m above sea level.

In Colchis very little of the original forest remains, and only in relatively inaccessible mountainous places. Samples from Colchic *Fagus* forest with evergreen understorey, from about 1,000 m on Mt. Mtirala near Batumi, are shown in Table 6.1. This very stately forest reaches 35 m in height. The forest, as represented by an area of 40 × 30m on a 30 ° mid-slope, is nevertheless an extremely simple forest, completely dominated by *Fagus orientalis* and containing a total of only nine species. The understorey is composed entirely of evergreen Colchic elements, mainly *Rhododendron ponticum* and *Ilex colchica* plus *Laurocerasus officinalis* and *Ruscus ponticus*. On a lower slope below the access road the forest was even taller, over 40 m, but contained *Rubus nigra* (sensu lato) and *Castanea sativa* in place of *Vaccinium arctostaphylos* and *Viburnum orientale*. One 40 × 40 m plot below the

Fig. 6.3 (continued)

road contained a total on only six species, with only *Rhododendron ponticum* as a significant understorey. All plots showed some seedlings of *Fagus* or Colchic evergreens (mainly *Rhododendron*) in the herb layer, suggesting that the forest is regenerating and will remain if not destroyed by man (Box et al. 2000).

An example of a montane *Fagus-Picea* forest from about 1,700 m on a steep north-facing slope in this area is shown in Table 6.2. *Fagus orientalis* is the dominant species in all but the herb layer. *Picea orientalis* also occurs in the canopy but did not seem to be regenerating significantly in the *Fagus* forest (Fig. 6.4). The herb layer is diverse, but the total cover (20 %) was low despite the relatively open canopy (60 %).

According to Dolukhanov (2010), the upper limit of the vertical distribution of beech depends on temperature regime and the amount of winter precipitation, protecting young beech trees from frosts, while the lower boundary depends on

Fig. 6.3 (**a**) Fagus orientalis, Dmanisi region, Lower Kartli. (Photo O. Abdaladze), (**b**) *Fagus orientalis*, Bakuriani (Photo Sh. Sikharulidze)

air humidity. Beech is the most shade- enduring plant among the deciduous woody species of Georgia.

One of the most characteristic species of the beech woods is *Carpinus caucasica* (=*C. betulus*). Under the continental climatic conditions, beech forests are replaced by hornbeam forest communities. In Colchis, *Castanea sativa* is a common associate of beech woods. Among the other species accompanying *Fagus orientalis* are: *Quercus iberica, Q. macranthera, Fraxinus excelsior, Ulmus elliptica, Acer platanoides, A. laetum, Tilia caucasica, T. cordata, Sorbus caucasigena,* etc. Examples of species present exclusively in Colchic beech forests are *Vaccinium arctostaphylos, Viburnum orientale, Trachystemon orientale*. All these species are the local endemics of Colchis. Although in the majority of cases the dominant position in forests is occupied by beech, it frequently is a constituent of the formation of beech-fir, beech-chestnut and beech-spruce forest communities.

Table 6.1 Colchic *Fagus* forest samples on Mt. Mtirala, near Batumi (relevés G-3, G-4, G-5) (Box et al. 2000)

	Relevé G-3 (1,120 m, 30° SSE)				Relevé G-4 (990 m, 20° SSE)				Relevé G-5 (990 m, 20° SSE)			
	T1	T2	S	H	T1	T2	S	H	T1	T2	S	H
	35 m	12 m	3 m	2 m	50 m	20 m	5 m	8 m	50 m	20 m	4 m	8 m
	85 %	10 %	75 %	15 %	80 %	20 %	40 %	30 %	80 %	10 %	80 %	5 %
Fagus orientalis	5.4	2.2			5.4	2.2	3.3	1.1	5.4	2.2	+.2	+
Castanea sativa							1.1					
Rhododendron ponticum			3.4	1.2			2.3	+			5.4	1.2
Laurocerasus officinalis			2.3				2.3				+	
Ilex colchica			3.3	1.1			1.2	+				
Vaccinium arctostaphylos			+.2									
Viburnum orientale			+									
Euonymus latifolia			+				+					
Ruscus ponticus			1.2	+			1.2	1.2				
Rubus hirtus							3.4	3.3			+.2	+.2
Hedera colchica		+	+	2.2				1.2				+.2

In the mountains populated by coniferous forests (1,200–2,000 m), beech woods have partially been extinct. Pure beech forests are not replaced by coniferous woods.

Beech woods occur on brown forest acid soils (dry and fresh forest types) and brown forest pseudopodzols (moist types of forest). Dolukhanov (2010) differentiates two classes of associations in the beech woods of Georgia: beech woods without developed woody undergrowth (underwood) and beech woods with Colchic woody undergrowth.

According to the above-mentioned author, the following groups of associations belong to the first class, i.e. that lacking shrubby undergrowth (some groups are not included):

1. Fageta festucosa (*Festuca drymeja*)
2. Fageta nuda
3. Fageta dentariosa (*Dentaria iberica*)
4. Fageta asperulosa (*Asperula odorata*)
5. Fageta pachyphragmosa (*Pachyphragma macrophyllum*)
6. Fageta rubosa (*Rubus hirtus*)

Table 6.2 Montane *Fagus-Picea* forest above Bakuriani (relevé G-8) (Box et al. 2000) (Location: Above Bakuriani, roadside slope below treeline (25 May 1999))

T1	25 m	60 %	1,700 m, Slope: 30° to NW	
T2	18 m	20 %		
S	3 m	10 %		
H	0.5 m	20 %		
M		1 %		
			30 × 30 m	AM, GN, KF, *EB*, RJL
T1	4.4	*Fagus orientalis*	2.2	*Picea orientalis*
T2	2.2	*Fagus orientalis*		
S	2.2	*Fagus orientalis*	+.2	*Picea orientalis*
	+	*Carpinus caucasica*		
H	2.2	*Dryopteris filix-mas*	2.3	*Arum albispathum*
	1.2	*Myosotis sylvatica*	1.2	*Asperula odorata*
	1.2	*Polygonatum verticillatum*	1.2	*Arabidopsis thaliana*
	1.2	*Paris quadrifolia*	+.2	*Petasites albus*
	+.2	*Urtica dioica*	+	*Rubus saxatilis*
	+	*Veronica peduncularis*	+	*Calamagrostis arundinacea*
	+	*Anthriscus nemorosa*	+	*Geum rivale*
	+	*Carex sylvatica*	+	*Campanula rapunculoides*
	+	*Asplenium nigrum*	+	*Ribes biebersteinii*
	+	*Senecio rhombifolius*	+	*Ligusticum alatum*
	+	*Senecio caucasica*	+	*Erodium cicutarium*
	+	*Taraxacum officinalis*	+	*Geranium sylvaticum*
	+	*Orobanche* sp.		
M	+	*Polytrichum commune*		

7. Fageta trachystemosa (*Trachystemon orientale*)
8. Fageta filicosa (*Dryopteris filix-mas, D. pseudomas,* etc.)
9. Fageta luzulosa (*Luzula sylvatica*)

Among beech woods with Colchic woody undergrowth the following may be mentioned:

1. Fageta rhododendrosa (*Rhododendron ponticum, Rh. ungernii*)
2. Fageta laurocerasosa (*Laurocerasus officinalis*)
3. Fageta ilicitosa (*Ilex colchica*)
4. Fageta ruscosa (*Ruscus colchicus* (Fig. 6.5))
5. Fageta magnovacciniosa (*Vaccinium arctostaphylos*)
6. Fageta azaleoza (*Rhododendron luteum*)
7. Fageta viburnosa (*Viburnum orientale*)

Below, we have attempted to give a brief description of some association groups of beech woods which is based on the results of investigations carried out by Dolukhanov (2010).

Fageta festucosa (Fig. 6.6a). These communities are widely distributed in East Georgia at 1,100–1,750 m above sea level. In the western part of the country they are replaced by fir and spruce forests and are a rare phenomenon here. *Carpinus*

Fig. 6.4 *Fagus orientalis*, *Picea orientalis* with *Paeonia steveniana* (Photo Sh. Sikharulidze)

caucasica is frequently presented in the first stratum along with the leading species of beech. These communities prefer relatively dry climatic conditions and good lighting. At comparatively higher altitudes, the above-mentioned forests are restricted to the slopes of southern exposure.

Fageta nuda. Bare-floor beech woods are confined chiefly to the mountains of the Greater Caucasus. They are poorly represented in Colchis. Bare-floor beech woods lack shrubby undergrowth, but at the same time not a single herb is present in the undergrowth. These forests are distributed at an altitude varying from 500–600 to 1,200 m, only rarely at 1,400 m a.s.l. The fact that Transcaucasian bare-floor beech forests are distributed mostly in Georgia is of particular interest. They are restricted to the slopes of northern exposure. Bare-floor beech forest in Georgia occurs on the soils characterized by poorly developed horizon of humus and belongs to the subtype of brown forest pseudopodzolic soils with a strongly

Fig. 6.5 *Ruscus colchicus*, South Caucasus (Photo D. Kharazishvili)

pronounced process of pseudopodzoliation. (Dolukhanov and Urushadze 1968). The Colchic variant of *Fageta nuda* reminds that of East Georgia. The main difference is in the presence of *Vaccinium arctostaphylos, Rhododendron luteum, Ilex colchica,* and *Laurocerasus officinalis* in the clearings of beech forests (although these species can rarely be found).

The origin of bare-floor beech forests is still under question.

Fageta asperulosa is a corresponding association to the west European beech woods with the same species (*Asperula odorata* = *Galium odoratum*) in the herbaceous undergrowth. But these Caucasian and European beech forest communities vary by their floristic composition and their process of succession; based on this difference, Dolukhanov (2010) is inclined to regard the Caucasian communities as a separate group – Fageta asperulosa caucasica. The woody undergrowth is missing here except for a small amount of *Ilex colchica.*

Beech forests with *Galium (Asperula) odoratum* are to be found in comparatively moist habitats, on well-drained brown forest soils. They are situated on the slopes of medium steepness of the northern exposure; their altitudinal distribution ranges from 1,100 to 1,550 m.

Fageta pachyphragmosa (Fig. 6.6b)**.** These communities require moist temperate conditions. The regeneration is rather high in these beech woods. The subdominant *Pachyphragma macrophyllum* belongs to an ancient, local endemic monotypic genus of Colchic stock. These beech woods are characterized by the admixture of such arboreal species as *Acer velutinum, A. pseudoplatanus, Tilia begoniifolia, Fraxinus excelsior,* sometimes *Abies nordmanniana. Ulmus glabra* (*U. elliptica*) was an associate of beech until the second half of twentieth century. The woody undergrowth consists of *Sambucus nigra* and *Corylus avellana*, although the latter is often absent. The herbaceous field layer is well developed.

Fig. 6.6 (continued)

Typical forests of this group are met at the altitudes from 500 to 1,000–1,100 m.

Fageta pachyphragmosa are well represented in Lagodekhi Reservation (East Georgia), where it is protected from destruction. The upper limit of such communities reaches 1,820 m in the mountains of Upper Svaneti (Dolukhanov 2010).

Fageta rubosa. These forests are confined to the middle parts of the forest belt; they are commonest among the moist beech woods of Georgia. These forests are characterized by a relatively high regeneration and dominance of *Rubus hirtus (sect. Glandulosi)* in the undergrowth. In Georgia, the area of distribution of Fageta rubosa comprises the southern slopes of the Greater Caucasus mountain system. The above communities are less characteristic of the Lesser Caucasus.

Typical Fageta rubosa are met on northern slopes at 1,100–1,600 m.

Fageta trachystemosa are characteristic for West Georgia. *Trachystemon orientale* is a relic species of Colchic stock. Under the moist climatic conditions

Fig. 6.6 (a) Fageta festucosa (Photo A. G. Dolukhanov), (**b**) Fagetum with *Pachiphragma macropyllum*, Lagodekhi National Park (Photo O. Abdaladze)

it reveals a high synecological amplitude from the sea level up to the alpine zone (700–1,200 m). In these forests, *Corylus* and *Sambucus nigra* are solitary, and the semi-prostrate *Viburnum orientale*, as well as other Colchic plants are to be found.

Fageta filicosa. These communities develop under moist conditions. One can find a stratum of large ferns, manifested by *Dryopteris filix-max, Athyrium filix-femina, Matteuccia struthiopteris,* etc., in this type of forest. It must be stated that natural regeneration is suppressed in the above-mentioned communities. Beech woods with ferns are widely distributed in West Georgia, whereas in the east of the country under relatively dry conditions, these forests are restricted to the upper parts of the forest belt. The floristic diversity is the most striking feature of beech forests with ferns.

Fageta luzulosa are typical only for upper parts of the forest belt of Svaneti (western part of the Greater Caucasus). Woody undergrowth consists of *Vaccinium arctostaphylos* with rather abundant *Laurocerasus officinalis, Rhododendron luteum, Viburnum orientale*. Herbaceous vegetation is sufficiently developed. Beech forests with Colchic undergrowth are of considerable interest because of to the absence of their analogues within the territory of West Eurasia. The Colchic undergrowth is rather dense and is constituted of semi-prostrate evergreen and partially deciduous shrubs. Floristic composition of the above undergrowth is related florogenetically to the relic complex of semi-prostrate plants, which represent subalpine crook-stem forests of Colchis. These plants require wet climatic conditions and equal distribution of precipitation during the year.

Fageta rhododendrosa unites two different associations of beech woods with *Rhododendron ponticum* (Fig. 6.7) and those with *R. ungernii*.

Fig. 6.7 Colchic forest with *Rhododendron ponticum* (Photo. G. Nakhutsrishvili)

The first association is one of the commonests in Colchis. It can be found almost everywhere in West Georgia, while in the eastern part of the country it is quite rare. Under the moist climatic conditions (in districts with the annual precipitation over 1,400–2,000 mm) it is distributed from the sea level to 1950 and even to 2,100–2,200 m. The most favourable conditions are provided at the altitude of 1,000 to 1,300–1,500 m above sea level.

Beech forests with *Rhododendron ungernii (Rh. ungernii* is a relic local endemic of Colchis) are typical for districts with a very high amount of precipitation (3,000 mm p.a.). They are to be found at an altitude of 1,200–2,000 m. Under very moist climatic conditions they descend to 800, sometimes to 600 m a.s.l.

Fageta laurocerasosa. These communities extend from the sea level up to 2,250 m. Typical beech woods with *Laurocerasus* are distributed from 700 to 2,000 m. Their vertical distribution depends on the humidity of climate. *Laurocerasus officinalis, Rhododendron ungernii*, require winter precipitation to protect them from frost. Unlike Fageta rhododendrosa, beech forests with *Laurocerasus* are well developed on limestones, and under the moist climatic conditions they are situated on the southern slopes with abundant sunlight. These communities occupy also some gorges of East Georgia.

Fageta ilicitosa. These communities extend from 500 up to approximately 2,000 m, but they are most common from 1,000 up to 1,800 m. They occupy a larger distribution range than beech woods with *Rhododendron* and they almost coincide with that of Fageta laurocerasosa. Beech woods with *Ilex* are typical for Colchis, though in some localities they penetrate into East Georgia. It must be noted, that *Ilex colchica* is more shade-enduring, than *Ilex aquifolium*.

Fageta magnovacciniosa comprises the most common associations of West Georgia. In East Georgia these communities are rare. In the west, they are very common from 900 up to 2,150 m above sea level. They never grow below 500 m.

Vaccinium arctostaphylos is a relict endemic species of Colchis closely allied to the plants distributed nowadays in Japan and Maccaronesia (Madeira Island). That species is rather shade-enduring and is connected with beech woods.

Typical beech-woods with *Vaccinium arctostaphylos* can be found in the areas where precipitation amount does not exceed 1,400–1,500 mm p.a.

Fageta azaleosa associations prefer both moist and relatively dry climate. On the mountains of West Georgia they extend from 800 up to 1,500 m above sea level, in the areas with precipitation amount of 1,200–2,000 mm p.a. Beech-woods with *Rhododendron luteum* of the upper part of forest belts extend to 1,900 m. These beech-woods in East Georgia are situated on prominent mountain slopes from 1,000 to 1,700 m above sea level (Aragvi river basin, Upper Alazani and Iori rivers).

Fageta viburnosa associations are characterized by a narrow synecological distribution range. *Viburnum orientale* is a local endemic of Colchis; mainly, it forms undergrowth in beech-woods, but it can also be found in fir forests. Beech-woods with *Viburnum orientale* occupy moist ecotopes of Colchis, though they can also be met in East Georgia (Aragvi river basin and Upper Alazani). The above mentioned communities extend on the northern slopes of 900–1,900 m above sea level. Typical beech-woods with *Viburnum orientale* are developed in a middle forest belt of West Georgia (1,100–1,600 m). These communities are peculiar for the presence of blackberry (*Rubus hirtus* group), as well as of *Hedera colchica, Trachystemon orientale* and *Dentaria bulbifera*. In the upper forest belt (above 1,700 m) beech-woods with *Viburnum orientale* are spread only in West Georgia.

6.3.2 Dark Coniferous Forests

In Georgia, 5–6 % of the entire forest area is occupied by coniferous forests (forests dominated by spruce) (5, 6 %) and fir (10 %). Excluding beech-woods, dark coniferous forests are most abundantly developed in Georgia.

Fir forests are dominated by the Caucasian fir, *Abies nordmanniana*, whereas the Caucasian spruce, *Picea orientalis,* is a leading species in spruce forests. Caucasian fir (belonging to *sect. Abies*) is closely allied to *Abies bornmuelleriana*, which inhabits northern Anatolia. *Picea orientalis* belongs to *sect. Omorika*; as a typical species of this section, *P. omorika* occurs on rocky limestones of South-West Serbia.

In these forests, the dominant species of fir and spruce are associated with oriental beech. *Pinus kochiana* is more frequent in spruce forests than in fir communities (Fig. 6.8a, b, c).

Dark coniferous forests are widely distributed in Western Georgia and in the western part of East Georgia. These forests never form a continuous belt, though their development is connected with the definite altitudinal-climatic zone, which extends from 1,000 up to 2,000 m a.s.l. Dark coniferous forests represent the commonest type of vegetation between 1,400 and 1,900 m.

Dark coniferous forests of Georgia and of the Caucasus, in general, are quite different from those of the taiga. Some typical representatives of taiga vegetation,

Fig. 6.8 (continued)

such as larch, boreal species of spruce and fir, though unavailable in the Caucasus, can be found in West Europe (Dolukhanov 2010).

The distribution area of oriental beech covers that of *Picea orientalis* (Fig. 6.9) and *Abies nordmanniana* (Fig. 6.10), while the distribution area of Caucasian fir coincides with that of the Caucasian spruce. However, due to the factor of moisture necessary for *Abies nordmanniana*, fir is missing in the most continental parts of its area. In East Georgia spruce penetrates into the area of upper parts of the river Didi Liakhvi and of the Aragvi gorge (in the Greater Caucasus). *Picea orientalis* occupies northern slopes of the Lesser Caucasus up to Tedzami basin; it is also scattered around the districts of Tbilisi (upper reaches of r. Vera). In dark coniferous forests of East Georgia the role of spruce is reduced.

Fig. 6.8 (**a**) *Pinus kochiana*, Tetrobi Reserve, Javakheti plateau, Lesser Caucasus, (**b**) Pine forest on the rocks, Kazbegi, Central Caucasus, (**c**) *Pinus kochiana*, Bakuriani, Trialeti range (Photo O. Abdaladze)

Fig. 6.9 Piceetum, Bakuriani, Trialeti range (Photo O. Abdaladze)

Abies/*Picea* forests occur on crystalline schists and rocky soils.

In fir dominated forests, *Fagus orientalis* is a common species.

Although dark coniferous forests are very rich in endemic and relic species, no exclusive species is to be found here.

Spruce and fir possibly penetrated into the Caucasus earlier than the Quaternary, in the Paleogene and mostly in the Pliocene. In the postglacial period spruce forests

Fig. 6.10 *Abies nordmanniana*, the Lesser Caucasus (Borjomi-Bakuriani area) (Photo O. Abdaladze)

withdrew far westwards to their present limits because of climatic changes; there were replaced by broad-leaved and pine forests. Later the pine forests were displaced by broad-leaved forests and their small fragments remained on steep rocky and stony slopes in some ravines and moraines (Gulisashvili 1949).

Reduction of temperature during some of the Pleistocene periods promoted maximal development and wide distribution of dark coniferous forests. According to Dolukhanov (2010), their further decline is due to fir and dark beetles. From the view point of Kolakovsky (1961), this phenomenon is due to the natural process of substitution of dark coniferous forests in Colchis by broad-leaved ones, namely beech woods, caused by the changes of climate during the Quaternary.

Floristic composition of fir and beech forests seems identical; boreal elements are more abundant in spruce forests.

Dolukhanov (2010) specifies two classes of associations of these forests on the basis of their undergrowth. The first class of dark coniferous forests is devoid of any developed woody undergrowth and includes the following groups of associations:

1. Piceeta orientalis maculato-muscosa
2. Abieteta festucosa, Fageto-Abieteta festucosa, Piceeta-Abieteta festucosa, Piceeta festucosa
3. Piceeta sicca
4. Abieteta luzulosa
5. Piceeta nanoherbosa, Piceeta-Abieteta nanoherbosa, Abieteta nanoherbosa, Fageto-Abieteta trachystemosa
6. Fageto-Abieteta filicosa
7. Abieteta subalpina calamagrostidosa
8. Abieteta subalpina heteroherbosa
9. Abieteta rariherbosa

Dark coniferous forests with Colchic undergrowth belong to the second class and include the following groups of associations:

1. Piceeta rhododendrosa, Piceeta-Abieteta rhododendrosa, Abieteta rhododendrosa, Fageto-Abieteta rhododendrosa
2. Piceeta laurocerasosa, Piceeta-Abieteta laurocerasosa, Abieteta laurocerasosa, Fageto-Abieteta laurocerasosa
3. Piceeta ilicitosa, Piceeta-Abieteta ilicitosa, Abieteta ilicitosa, Fageto-Abieteta ilicitosa
4. Piceeta-Abieteta magnovacciniosa, Abieteta magnovacciniosa, Fageto-Abieteta magnovacciniosa
5. Abieteta viburnosa, Fageto-Abieteta viburnosa

Piceeta orientalis maculato-muscosa. Associations of this group are characteristic for spruce and pine forests. They extend from 1,100 to 2,100 m a.s.l. The characteristic and constant species is *Goodyera repens*. Within well developed bryophytes in these forests the following may be mentioned: *Hylocomium splendens, Pleurozium schreberi,* etc. The appearance of these forests as well as the floristic composition of bryophytes is very similar to those of Eurasia boreal zone. Mossy spruce forests are well represented in the western part of East Georgia and in Colchis (Nakhutsrishvili, Hübl et al. in press).

Dark coniferous forests with *Festuca*. The predominance of *Festuca drymeja* in the herbaceous undergrowth is characteristic for fir, fir-spruce and beech-fir forests. The altitudinal distribution of these forests ranges from 900 to 2,100 m. The above-mentioned communities are characterized by the presence of following species: *Dryopteris filix-mas, Oxalis acetosella, Sanicula europaea, Viola reichenbachiana.*

Piceeta sicca. These communities are spread only in East Georgia at 1,000–1,500–1,700 m. They occur on poor skeletal soils of steep and sunny slopes. *Picea orientalis* is associated here with pine and Georgian oak (*Quercus iberica*). The ground layer is rather poor in species (*Poa nemoralis, Brachypodium sylvaticum, Oxalis acetosella,* etc.).

Abieteta luzulosa (*Luzula sylvatica*). These forests are restricted to the mountains of Svaneti (West Georgia). They extend from 1,600 to 2,150 m. These communities populate almost all the slopes, though northern slopes are still more favourable for them. There is no woody undergrowth in these forests, although the ground layer is well developed.

Dark coniferous forests with dwarf herbs. These associations prefer moist climatic conditions; they extend from 1,300 up to 2,000 m nearly in all massifs of dark coniferous forests. The main components of these communities are dwarf herbs; grasses are rather poor in species. In East Georgia these communities are exemplified by spruce forests, in the west of the country, however, by fir, fir-spruce and beech-fir forests. Among the species characteristic for fir and spruce communities the following may be mentioned: *Oxalis acetosella, Sanicula europaea,* and *Galium rotundifolium*. Frequently occurring as underwood in spruce forests of West Georgia are *Vaccinium arctostaphylos* and *Daphne pontica*.

Piceeta trachystemosa. These communities occur chiefly in the mountains of northern Colchis (900–1,900 m a.s.l.) under slightly moist conditions. *Fagus orientalis* is a common associate of spruce in these forests; sometimes spruce is even replaced by beech. Underwood consists of scattered *Vaccinium arctostaphylos, Viburnum orientale, Ilex colchica,* and *Rhododendron ponticum*. The ground layer is characterized by the prevalence of *Trachystemon orientale*.

Fageto-Abieteta filicicosa. These communities occupy almost all massifs of fir forests of West Georgia. They belong to the middle and upper parts of dark coniferous forest zone and are characterized by the presence of large ferns: *Athyrium filix-femina, Dryopteris filix-mas, D. oreades, Oreopteris limbosperma, Matteuccia struthiopteris,* etc. Natural regeneration is not satisfactory.

Dark coniferous forests with Colchic undergrowth are not as widely distributed as those of beech. As it was pointed out above, they occupy the mountains of West Georgia.

A special attention should be paid to the series of dark coniferous forests with *Rhododendron ponticum*; spruce, fir and beech are arboreal components of these communities. *Rhododendron ponticum* is more abundant in beech-spruce forests. Such forests can be rarely met in East Georgia. Dark coniferous forests above 1,900 m can be found only in the mountains with moist climate (Dolukhanov 2010).

Natural regeneration is suppressed here. Dark coniferous forests with *Laurocerasus* are peculiar for the upper part of the dark coniferous forest belt. They are developed under moist climatic conditions, but, contrary to those with *Rhododendron ponticum*, they are restricted to the slopes of southern exposure; they also occur on limestones. Fir forests with *Laurocerasus* represent a very rare type of vegetation in East Georgia.

Table 6.3 Chirukhistkali gorge, Mt. Chirukhi (Kharazishvili 2005)

T1 22–25 m	70%	1,990 m, 20 ° N
T2 10–12 m	20%	
S 2.5 m	60%	
H 2 m	80%	
T1	*Abies nordmanniana*	4
	Picea orientalis	3
	Acer trautvetteri	+
T2	*Abies nordmanniana*	2
	Picea orientalis	+
	Acer trautvetteri	+
S	*Vacciniumm arctostaphylos*	3
	Lonicera caucasica	2
	Rubus buschii	2
	Viburnum opulus	+
	Daphne pontica	+
H	*Senecio propinquus*	3
	Athyrium filix femina	3
	Pyrola minor	3
	Valeriana alliariifolia	3
	Festuca montana	2
	Lilium kesselringianum	1
	L. szowitsianum	1
	Taraxacum litwinowii	1
	Dactylorhiza flavescens	1
	Gadellia lactiflora	1
	Silene wallichiana	+
	Geranium psilostemon	+
	Veratrum lobelianum	+
	Polygonatum verticillatum	+
	Trifolium pratense	+
	Alchemilla retinervis	+
	Plantago lanceolata	+
	Prunella vulgaris	+
	Pyrethrum roseum	+
	Ranunculus repens	+
	Luzula forsterii	+
	Fragaria vesca	+
	Tussilago farfara	+
	Petasites albus	+
	Hesperis matronalis	+
	Grossheimia polyphylla	+
	Knautia involucrata	+
	Galium album	+
	Centaurea nigrofimbria	+

(continued)

Table 6.3 (continued)

	Scrophularia chlorantha	+
	Potentilla recta	+
	Heracleum cyclocarpum	+
	Pyrethrum marcophyllum	+
	Aruncus vulgare	+
	Sanicula europaea	+
	Gentiana shistocalyx	+
	Rumex acetosella	+
	Carex capitellata	+
	Cirsium kosmelii	+
	Sedum tenellum	+
	Symphytum asperum	+
	Paris incompleta	+
	Inula orientalis	+
	Oxalis acetosella	+

Laurocerasus officinalis is less shade-resistant than *Rhododendron ponticum.* That is why it is not typical for dense fir forests. In the underwood of dark coniferous forests, *Ilex colchica* can be frequently observed with *Laurocerasus*; in many cases the above-mentioned species occupies a dominant position and forms series of associations. Among these associations, restricted to the middle part of the forest zone (1,200–1,700 m), special attention is paid to *Picea* forests with *Ilex* and with *Fagus-Ilex. Vaccinium arctostaphylos* and *Ruscus colchicus* are rather rare in Georgia. *V. arctostaphylos* is very common in the second stratum of beech-fir forests from 1,200 to 1,900 m. The ground vegetation is composed of *Festuca drymeja, Asperula (Galium) odorata, Oxalis acetosella, Dentaria bulbifera,* etc.

Associations of spruce-fir forests with *Vaccinium arctostaphylos* occur only in Svaneti. The less common group of associations is Piceeta viburnosa. These communities are confined to several parts of Svaneti and extreme western parts of Trialeti Mts.

One of the examples of the association **Abieteta subalpina heteroherbosa** is represented in the river of Chirukhistskali gorge, Adjara (Colchis) (Kharazishvili 2005). The community is distinguished by diversity of shrubs as well as herbs (Table 6.3). Undergrowth is constituted by such Colchic elements as *Daphne pontica* (Fig. 6.11), *Vaccinium arctostaphylos.* The following species of the subalpine meadows are worth mentioning: *Pyrethrum roseum, Alchemilla retinervis, Ranunculus ampelophyllus, Lilium spp., Geranium psilostemon, Dactylorhiza flavescens*, etc. Representatives of the tall herbaceous vegetation occurring in coenoses of the association are *Senecio propinquus, Athyrium filix femina, Grossheimia polyphylla, Heracleum cyclocarpum, Gentiana schistocalyx.* The following endemics can be found in this community: *Senecio propinquus, Gadellia lactiflora, Symphytum asperum* (Caucasian), *Grossheimia polyphylla, Lilium kesselringianum* (Colchic).

Fig. 6.11 *Daphne pontica*, Colchis (Photo Z. Manvelidze)

The forest owing to the mountain resort Bakuriani on the Lesser Caucasus is grazed by cattle (Nakhutsrishvili, Hübl et al. in press, s. Table 6.4). The tree layer is made up of *Picea orientalis*, *Pinus kochiana* (*Pinus sylvestris* subsp. *kochiana*) and *Fagus orientalis*. Grazing creates a mosaic structure consisting of forest with typical undergrowth and thinnings with plants typical of pastures. The study was conducted to find out vegetation units of the grazed forest. Twenty one relevés were compiled according Braun-Blanquet scheme. It was impossible to separate forest and pastures by micro-relevés. Therefore most of the descriptions contain portions of both forest and pasture.

The common knowledge about the relation between the dominant trees *Picea orientalis*, *Pinus kochiana* and *Fagus orientalis* seems to be similar as in case of *Picea abies*, *Pinus sylvestris* and *Fagus sylvatica* in Central Europe. *Pinus kochiana* has the lowest demand for depth and moisture of the soil, but the highest on light. *Fagus orientalis* has the highest demand for soil and the lowest on light. *Picea orientalis* is in between.

Fagus orientalis dominates alone only in one relevé (1), and here the tree layer has the highest coverage (95 %). There are no lower trees and shrubs in the relevé. The herb layer is very poor (5 %). Plants typical of pastures are not present. The moss cover is also very poor (2 %). In the relevé 2 *Fagus orientalis* and *Picea orientalis* are co-dominants. The tree layer is the highest with 90 % coverage. The second layer (5 %) has a very low coverage and consists only of young *Fagus* and *Corylus avellana* trees. The herb layer is a little more developed, but also without plants of pastures. The moss cover is absent.

Picea orientalis is the most frequent tree and also the most frequently dominating, and reaches 80 % in the tree cover. *P. orientalis* is also the most frequent species in the layer of lower trees and shrubs, in 16 of the 21 relevés. *Fagus*

Table 6.4 Table of relevés within the Bakuriani Alpine Botanical Garden

Number of relevé	1	2	15	14	13	18	12	19	16	17	4	9	5	6	10	11	3	7	8	20	21
Fall	5	10	7	5	5	10	20	5	20	15	20	15	5	5	10	10	10	15	20	25	30
Exposure	S	SSE	SW	S	SW	SE	SW	S	SSW	SW	WSW	SW	W	E	SW	W	SW	SW	W	S	SSW
Plot size (m^2)	200	200	400	400	200	200	200	100	200	200	200	300	100	200	200	200	400	100	80	200	200
Height tree	25	30	30	30	30	30	30	30	30	35	30	30	30	30	30	30	30	30	30		30
Tree cover (%)	95	90	30	10	60	50	40	80	80	80	70	60	60	50	50	50	20	25	10		5
Height shrub		4	10	20	20	15	15	10	5	10	10	4	4	10	10	20	15	3	10	20	20
Shrub cover (%)		5	15	30	10	15	40	10	20	10	5	5	1	10	10	10	5	1	10	40	20
Hight herb	10	40	100	100	50	50	100	50	50	30	95	30	60	30	80	120	40	30	30	100	100
Herb cover (%)	30	10	70	50	60	60	50	15	2	10	40	35	10	60	50	70	10	5	60	50	60
Trees																					
Fagus orientalis	5	3	1	+																	
Picea orientalis		3	2	2	4	3	2	5	5	5	4	4	4	3	3	3	3	2	2		
Pinus kochiana			2	+	+	2	2														1
Pyrus caucasica	+			+																	
Shrubs																					
Fagus orientalis		+			+			1	+	+	1	+		1	1	1	+	+	+		
Picea orientalis			+	+	2	2	3	2	2	2	2	+	+	2	2	2			2		+
Pyrus caucasica			2	2	+	+															
Pinus kochiana						+													+	3	2
Lonicera caucasica			+		+	+		+						+			+			+	
Corylus avellana		+	+																		
Salix caprea			+											+	+						
Malus orientalis				+		+	+													+	
Prunus divaricata			+	+																	
Rosa canina agg.						1						+								+	+
Quercus macranthera						+														+	+

(continued)

Table 6.4 (continued)

Prunus avium						+															+
Ribes biebersteinii								+			+										
Herbs																					
Festuca drymeia	+	1	1		2		1	+	+	1	1	2	+	2	3	3	+	+	3		+
Fragaria vesca	+	+	1	1	2	+	+	+	+		1	+		1	+	+	2	+	2		1
Fagus orientalis	+		+			r	+			+	+	+	1	+	+			+			
Galium odoratum	+	+	+	+							1	1		1	+	2					
Orobus cyaneus		+	+	+	+	+	+	+	+	2	+	2		2	1	+	1		+		+
Sanicula europaea		1	+	+	1	+	+	+		+	1	+		1	+				+		
Primula macrocalyx		+	1	+	+	1	+	1			1	+		+	+	+	1		+	+	1
Primula woronowii		+	2	2	2	2	1	1			2	1		2	1	2	2		1		+
Picea orientalis		+	+				+	+	+			+	1	+							+
Prunella vulgaris			2	2	2	+	+				+	1		1	+	1	1		2	+	
Brachypodium sylvaticum			2	+	2	2	2	1													
Clinopodium vulgare			+	+	+	+	1			+	1	+		1	+	+	+		+		1
Trifolium ambiguum			1	1	+	+	2	+			+				+		+		+	+	
Valeriana tiliifolia			+	+	+	+	+	+			+	1	+	1	+		+	+	+		+
Digitalis ferruginea			+	+	+		+						+	+	+		+		+		+
Sedum stoloniferum			2	1			+	1			+	+	+	+	+	+	1				
Centaurea salicifolia			1	+	+	+											+		+		
Geranium gracile	+		1	1	+							+									
Luzula sylvatica			+		+		+	1			1			1	+	+	1	+	+		
Trifolium medium			+		+						+	+		+			+				
Poa nemoralis			+	+		+	1	1			1	1	+	1		1	+		+		1
Veronica peduncularis			1				+				+	+			+		+			+	+
Salvia glutinosa			+		+			+			+	+		+		+					
Ranunculus caucasicus			+	+	+	+	+														

Carex muricata agg.	+	1	+		+													+	
Pyrus caucasica	+	+		r									+		r			r	
Gentiana septemfida	+	+	+	+															
Hieracium hoppeanum	+			2	+							+	+						
Trifolium canescens	1				+							+				+			
Myosotis arvensis	+			+	+													+	+
Viola reichenbachiana	+				+					+			+	+	+				
Festuca pratensis		1		+					1	2				3					
Leontodon danubialis		2	+	1	+														
Leontodon hispidus		1	1	1	2													+	
Cruciata glabra		1		+	+										+			+	
Carex sylvatica		2	+	1			+		+										
Cardamine impatiens		+	+		+			+	+	+	+	+	+	1	+				
Lapsana grandiflora		+	+	+	+				+		+	1	+	+	1				+
Trifolium fontanum		+		+	+	+									+		1	+	
Achillea millefolium agg.		+		+					+						+			1	+
Gentiana schistocalyx		+	1			+								+			+		
Pyrethrum demetrii		+		1	+				+	+								+	
Mycelis muralis		+				+		+	+	+	+	+	+			+	+		
Moehringia trinervia		+				+			+			+	+	+					+
Sorbus caucasigena		+				+	+	+						+					
Alliaria petiolata		+							1					+	+				
Trifolium repens		+				+								+	1				
Urtica dioica		+				r								+	+				
Hieracium murorum			+		+	+	+	+	+	+	+	+	1		+	+	+		
Veronica officinalis			+		+	+		+	+	+		+	+			+	+		
Campanula rapunculoides			+		+	+			+	+	+	+	+				+	+	+

(continued)

Table 6.4 (continued)

Oxalis acetosella					+			+		1		1		1	2	+				
Potentilla crantzii					+	+	+												1	1
Dryopteris filix-mas					+						+		+	+	+					
Coronilla varia					+								+			+			+	+
Trifolium trichocephalum						+								+		+			1	1
Silene ruprechtii				r		+	+	1												1
Myosotis sylvatica							+					+	+		+	+			+	
Calamintha grandiflora							+	+		+		+		+	1					
Plantago lanceolata							+									+		+	+	+
Galium aparine								+	r		+	+	+	+		+				
Galega orientalis								+			+	1	+	+						
Orthilia secunda								+	+	1							+			
Solidago virgaurea								+		+	+	+		+						
Geranium robertianum				+							+	+			+	+				
Lonicera caucasica					r			+				+			r					
Lamium album											+		+			+				+
Cruciata laevipes											+		+			+				+
Cytisus caucasicus													+	+				+	1	1
Dentaria bulbifera	+				r								+							
Prunus avium	+															+				+
Polygonatum multiflorum	+	+													+					
Lathyrus pratensis			+			+								r						
Ajuga genevensis				+		+												+		
Hieracium vulgatum				+			+							+			+			
Trisetum flavescens				1												+			+	
Viola tricolor agg.				+												+			+	
Quercus macranthera						r	+	+												

Sedum gracile						+	+												+
Astrantia trifida						1								r		+			
Corylus avellana						+			+			+							
Chaerophyllum aureum						+				+						+			
Dactylorhiza urvilleana						+				+							+		
Cirsium cosmelii							+							r					+
Cerastium holosteoides							+								+				+
Rubus caucasicus									+				+	r					
Anemone caucasica	+	+																	
Vicia			+										+						
Prunus divaricata			+															+	
Silene italica			+															+	
Lilium szovitsianum				r	r														
Alchemilla rigida				+		+													
Cynosurus cristatus				2		+													
Poa trivialis				+											1				
Rumex obtusifolius				+											+				
Sedum tenellum				+														1	
Geranium pallens					+						+								
Taraxacum officinale agg.						+					r								
Neottia nidus-avis							+	+											
Astragalus glycyphyllos							2			+									
Sambucus ebulus								+		+									
Goodyera repens									+					+					
Cicerbita racemosa										1				+					
Galium valantioides										+						+			
Cephalanthera rubra												+	1						

(continued)

Table 6.4 (continued)

Sedum pallidum																	+				+
Veronica chamaedrys																			+		+
Coronilla balansae							1													1	1
Arenaria serpyllifolia																				1	+
Alyssum murale																				1	1
Cephalaria procera																				+	1
Carex humilis																				1	+
Helianthemum grandiflorum																				1	2
Melampyrum arvense																				+	+
Phleum phleoides																				1	+
Origanum vulgare																				+	1
Scabiosa bipinnata																				1	+
Teucrium nuchense																				+	+
Verbascum pyramidatum																				+	+
Number of taxa	13	18	44	62	46	55	60	43	12	21	50	41	17	47	45	44	50	15	30	54	61
Moss cover (%)	2	0	10	5	80	20	10	70	20	40	80	80	80	70	60	40	25	90	70	5	0
Hylocomium splendens								3			4	3	4		3			5			
Pleurozium schreberi								3				3	3		1			1			
Mnium sp.															+						

orientalis occurs in this layer in 13 relevés. In the herb layer *Fagus* is a little more frequent (in 11 relevés) than *Picea* (in nine relevés). This is remarkable because *Fagus* was found only in five relevés in the highest tree layer. Therefore fruiting trees are not common. It seems that animals spread the fruits of *Fagus*. The low demand for light of the seedlings ensure there survival. Probably the young beech trees are eaten by cattle. This may explain the minor frequency and the minor coverage of *Fagus* in the second layer. Sometimes the reproduction of *P. orientalis* on mouldering trunks is observed, which is characteristic also to *P. abies*.

Pinus kochiana is the co-dominant with *Picea orientalis* in three relevés. In these cases the whole tree layer has a relatively low coverage (30 %, 40 %, 50 % in the relevés 12, 15, 18). The herb layer is well developed (50 %, 70 %, 60 %) and do not differ significantly from the relevés without *Pinus*.

The relevés 20 and 21 were made on a steep south facing slope with a loose tree layer of *Pinus kochiana* together with a few *Quercus macranthera*. The herb layer is well developed (50 % and 60 %) and differs markedly from the other relevés. In the both relevés species of open habitats prevail. *P. kochiana* do not occur in any relevé in the herb layer. The following Vasilevich and Bohn (2003) are islands of the forests of *P. kochiana* on steep slopes. Pine does not form a separate altitudinal zone.

The undergrowth is very diverse. In the relevés the species typical of forests prevail. The most common are *Festuca drymeja* and *Fragaria vesca* in 18 of 21 relevés. The second position has *Orobus* (*Lathyrus*) *cyaneus* (in 12 relevés) typical of pastures and meadows. A second relatively common species typical of meadows is *Trifolium ambiguum* (in 11 relevés). Several species occur in forests as well as on meadows, for instance *Prunella vulgaris* and *Poa nemoralis*.

Species which prefer substrate reach in bases like *Sanicula europaea* and *Galium odoratum* can be found as well as species growing on acid soil like *Luzula sylvatica* and *Veronica officinalis*. In spite of the fact, that *Picea orientalis* is the dominant tree, species of coniferous forests are rare. *Orthilia secunda* occurs in four relevés, *Goodyera repens*, which is typical of spruce forests in the Caucasus – only in 2. More frequent are mosses typical of coniferous forests: *Hylocomium splendens* (in six relevés), *Pleurozium schreberi* (in five relevés). It seems that the predominance of *Picea orientalis* is encouraged by human impact.

Many species of the herb layer occur also in Central Europe. Frequent species which are absent in Central Europe are *Orobus cyaneus*, *Primula macrocalyx*, *Primula woronowii*, *Valeriana tiliifolia*, *Trifolium ambiguum*.

Festuca drymeja which is most frequent and often dominating in the herb layer reaches the eastern border of the Alps in the Southeast in Slovenia and in the Northeast in Austria (Aeschimann et al. 2004). In the Vienna Wood *Festuca drymeja* is a characteristic species of the beech forests (Nakhutsrishvili, Hübl et al. Table 6.4).

6.4 Mixed Broad-Leaved Forests of Colchis

These forest communities belong to the Euxinian broad-leaved forests; they are restricted to Colchis. Stands of these forests can be observed on the eastern slopes of the Ajara-Imeretian mountain range, in the north-western part of the Greater

Caucasus, etc. Mixed forests of Colchis are extended from approximately 200 m up to 1,000–2,000 m, though in South Colchis they almost approach the sea level. These forests occur on podzolized zheltozems, as well as on mountain forest brown soils and krasnozems.

These forest communities are developed under moist climatic conditions with precipitation amount of 2,500 mm p. a.

Mixed broad-leaved forests are very rich floristically; a considerable number of relic mesophytic species of the Caucasus occur here. These forests mostly cover moist gorges with uniform air humidity all year long. High air humidity of these gorges is caused by the peculiarities of atmospheric precipitation and moderate temperature regime, which keep the surface of the slopes cool. Owing to these favourable conditions, many representatives of the Tertiary flora continue to exist in Colchis. A vivid example of Tertiary relics is the poikilohydric fern, *Hymenophyllum tunbridgense*, restricted to one of the moist gorges of South Colchis. Mixed broad-leaved forests are characterized by the occurrence of 50 arboreal and 80 herbaceous species (Kolakovsky 1961; Sakhokia 1980). Major forest-building species include: Old-Mediterranean *Castanea sativa*, Old-East-Mediterranean *Fagus orientalis*, Colchic-Hyrcanian *Zelkova carpinifolia*, Colchic *Quercus imeretina*, Euxinian *Q. hartwissiana*, south-Caucasian *Q. iberica* and Minor Asia-Caucasian *Carpinus caucasica*. The arboreal associates of the above-mentioned species are Colchic-Hyrcanian *Acer laetum* and *Pterocarya pterocarpa*, Colchic *Ficus colchica* and *Salix alba* subsp. *micans*, Caucasian *Pyrus caucasica, Malus orientalis* and *Salix pantosericea*, as well as European *Acer platanoides, Fraxinus excelsior, Taxus baccata,* etc. Additionally, a mention should be made of *Tilia caucasica, Ulmus glabra, U. elliptica;* Mediterranean-East-Asian *Diospyros lotus,* etc. Mediterranean *Rhododendron ponticum, Ruscus ponticus, R. colchicus, Daphne pontica;* Colchic *Ilex colchica, Rhododendron ungernii, Epigaea gaultherioides* and *Buxus colchica* frequently occur as undergrowth in these communities. Among the deciduous plants of the Colchic undergrowth, relics occupy the dominant position; examples are Colchic *Vaccinium arctostaphylos, Staphylea colchica, Viburnum orientale, Euonymus leiophloea, Hypericum xylosteifolium;* Colchic-Caucasian *Rubus caucasicus;* Colchic-Hyrcanian *Crataegus microphylla,* etc.

The ground vegetation (field layer) is characterized by the presence of such ferns as *Matteuccia struthiopteris, Athyrium filix-femina,* etc. Epiphytic ferns are exemplified by *Polypodium serratum.* Moist rock crevices and stony sites are inhabited by *Phyllitis scolopendrium, Pteris cretica,* etc.

Lianes (vines), though typical for Colchis forests, never form thickets even in forest margins. The commonest lianas are: *Hedera colchica* and *Dioscorea caucasica* (both are Colchic endemics); *Tamus communis* and *Periploca graeca* (Mediterranean species); *Smilax excelsa* and *Clematis vitalba,* etc. As indicated before, Colchic forests include epiphytes. Epiphytic plants are manifested by lichens (*Usnea barbata*), mosses (Neckeraceae representatives) and ferns.

Of the main plant communities *Fagus-Castanea, Carpinus-Castanea, Carpinus-Fagus-Castanea* and *Alnus-Carpinus-Fagus-Castanea* forests should be mentioned. Among these communities, forests dominated by five and more species can be observed. An example is the community where the dominant position is occupied by *Castanea sativa, Fagus orientalis, Tilia caucasica, Carpinus caucasica, Alnus barbata* and *Taxus baccata*. In the ground vegetation (where developed) *Brachypodium sylvaticum* and the adventive grass *Oplismenus undulatifolius* are dominants. They are associated with *Cardamine impatiens* and *Oxalis corniculata*.

Quercus imeretina forests with *Carpinus caucasica* are spread on river terraces.

It should be stated, that Colchic river gorges are characterized by a considerable amount of heterogeneity of natural conditions. Forest communities with *Buxus colchica* as undergrowth are typical for limestone regions, while in non-limestone areas forest undergrowth is composed of *Rhododendron ponticum, Rh. ungernii*, etc.

According to Kolakovsky, these moist gorges covered by protective heavy snow in winter, have prevented Tertiary flora remnants from destruction, which overwhelmed the similar vegetation of exposed slopes.

The term "Shkeriani" is usually applied to define thickets of evergreen shrubs and even creeping trees. Golitsin (1939) attributes this term to a whole complex of evergreen shrubs, including *Epigaea gaultherioides, Ilex colchica* and also to deciduous *Betula medwedewii, Quercus pontica, Vaccinium arctostaphylos, Viburnum orientale*. According to Kolakovsky, "Shkeriani" should be referred to the formation of shrubs, developed below the forest canopy and which now build up the second stratum. Mainly, the components of "Shkeriani" are the derivatives of forest vegetation, and such plants as *Epigaea gaultherioides, Ilex colchica, Betula medwedewii* should be excluded from its composition.

Destruction of Colchic forests in the lower part of the forest belt is the effect of man's agricultural activities; these areas, made subject to intensive farming, are manifested by the following cultivated plants: *Citrus* species, tea, tung tree, tobacco, maize, etc.

These areas are populated by many adventive plants like *Baccharis halimifolia* (North American), *Paspalum paspaloides* (pantropical), *Andropogon virginicus*, etc.

6.5 Pine Forests

Mountain pine forests of Georgia are dominated by *Pinus kochiana* which belongs to *P. sylvestris* group. The distribution area of *P. kochiana* comprises the Caucasus, Crimea and the northern Asia Minor up to the western part of the Pontic Mts. In the west, this species occurs in the Balkan Peninsula Tumajanov (1980) indicates that *P. kochiana* forests are to be found in all forest regions of moist temperate climate throughout the Euxinian phytogeographical province.

Main massifs hold by the above-mentioned forests, are situated on the mountains of the Greater Caucasus. In Georgia, pine forests cover negligible areas. In the eastern part of the country, these forests are distributed over the Mtkvari (= Kura) river basin, eastwards from Tbilisi and in Tusheti. In West Georgia they usually occur on stony and rocky slopes, where they cannot be replaced either by dark coniferous woods or by broad-leaved forest communities (Dolukhanov 2010).

Dolukhanov (2010) noticed an interesting exception, specific for the pine forests of Georgia. In some parts of East Georgia, namely in the depth of Tushetis and Pirikitis Alazani river gorges, pine forests typical for stony and rocky slopes, also occur in habitats with a developed soil cover. These biotopes are favourable not only for pine but also for such forest-building trees as spruce and beech. Nevertheless, in the gorges, mentioned earlier, spruce and beech are absent and the climax communities of pine forests are formed instead. Dolukhanov explains this phenomenon by the orographic and geographic isolation of these gorges. Due to historical circumstances, only pine has been preserved in these sites; it escaped being replaced by beech and spruce only because their seeds never got to this area.

Mountain pine forests extend from 700 up to about 2,400 m a.s.l. The most favourable conditions for pine forests are provided at an altitude of 1,000–2,200 m. In several parts, pines have been recorded at 2,500–2,600 m (Tumajaniv 1980).

P. kochiana grows on both limestone and acid magmatic rocks.

It is accepted that distribution of the pine forests of the Caucasus is closely connected with the regions of maximum glaciation in the Quaternary (Gulisashvili 1949). Formation of the present montane pine forests of the Caucasus took place under the influence of the Pleistocene glaciers (Tumajaniv 1980).

The main part of the pine forests in confined to the mountains of the Greater Caucasus and mostly, to its northern slope.

In the North Caucasus the pine forests form a wide line from 900–1,000 to 2,500–2,600 m a.s.l. On northern slopes in conditions of the deep snow cover they are usually replaced by birch forests (Betuletum) (Tumajaniv 1980).

Pinetum siccum astragalosum microcephalus is characteristic to the Lesser Caucasus.

Pine forests of the western South Caucasus are, in contrast, characterized by their connections with dark coniferous forests, especially spruce forests and formation of a spruce-pine association of the order Hylocomiosa.

Following Dolukhanov (2010), the distribution of *P. kochiana* is controlled by phytogeographical conditions, while the influence of climate is less important. Occurrence of climax communities of pine forests on stony and rocky slopes accounts for their floristic composition, which is rich in endemic species.

Within the pine forests of Georgia, the following groups can be outlined: climax communities of the above forests and temporal-derivative pine forests.

Pine is well known as a pioneer tree to occupy the naked stony substrata, morain and fluvioglacial drifts. In Transcaucasia and particularly in Georgia, development of pine forests is subject to the impact of climatic fluctuations and activities of man (Dolukhanov 2010).

The climax pine forests of Tusheti are very diverse, despite the fact that they occupy a limited area. They occur on substrata different by the character of weathering, as well as by the peculiarities of the lithological composition of the rocks (Tumajanov 1938; Dolukhanov 2010).

Dolukhanov (2010) differentiates the following categories of mountain pine forests:

1. Caucasian mountain pine forests of relatively dry ecotopes;
2. Caucasian mountain pine forests of moderate-moist ecotopes;

3. Caucasian mountain pine forests of swampy habitats;
4. Caucasian mountain pine forests occurring on stony screes and rock streams.

6.5.1 Caucasian Mountain Pine Forests Restricted to Rocky Slopes

Mountain pine forests of relatively dry ecotopes can be found almost everywhere within the distribution area of *Pinus kochiana*. They occur on skeletal and poorly-developed soils of prominent southward slopes. These pine forests are characterized by rich and diverse ground vegetation.

From the viewpoint of Dolukhanov (2010), several groups of associations can be distinguished within this forest type. Our aim is to give a short description of some of them.

Pine forests with sparse herbaceous undergrowth are confined to prominent southern exposure. Availability of scattered herbs in the undergrowth is due to the washout of soils and destruction of ground vegetation. Pine forests of dry habitats with xerophytic herbs in the undergrowth are usually the climax communities which occupy northern slopes of Trialeti Mts, Meskheti, etc. The second stratum of these forests consists of *Quercus iberica* (below 1,500 m), *Q. macranthera* (above 1,500 m) and species of *Acer*. Species dominant in the herbaceous undergrowth include *Carex buschiorum, Poa nemoralis, Brachypodium sylvaticum, Sesleria anatolica,* etc. Particular interest arise pine forests with tragacanthic species of *Astragalus* and those with *Juniperus (J. oblonga, J. hemisphaerica)* as undergrowth, frequent at 1,900–2,000 m a.s.l. Northern slopes of Trialeti Mts. are inhabited by pine forests with *Chamaecytisus caucasicus*, whereas those with *Ch. hirsutissimus* are restricted to mountainous Abkhazeti. In Ajara, the undergrowth of pine forests is dominated by *Cistus salviifolius*.

It is pertinent to note that Caucasian mountain-steppe pine forests and mountain meadow-steppe *P. kochiana* communities belong to the above-mentioned ecological type; they participate in generating park-like landscapes on the background of mountain meadow-steppe vegetation (usually in distant parts of the Javakheti Upland plateau). The communities mentioned earlier are characterized by rather dense ground vegetation, extremely rich floristically. These forests occur at the altitude between 1,700 and 2,400 m. These communities are also observed in Turkey.

Caucasian mountain pine forests of moderately moist ecotopes are characterized by the admixture of spruce, as well as *Betula litwinowii, Abies nordmanniana, Sorbus caucasigena, Fagus orientalis, Fraxinus excelsior,* etc.

According to Tumajanov (1938), the above-mentioned pine forests can be divided into the following series: (1) *Hylocomiosa*, including Pineta myrtillosa, Pineta oxalidosa, Pineta vacciniosa, and (2) *Composita* with Pineta rhododendrosa (*Rhododendeon caucasicum*), Pineta mixtofruticosa, Pinetum azaleosum, and Pinetum tiliosum communities.

As indicated above, these pine forests are examples of climax communities. The series Hylocomiosa deserves special interest. These communities are dominated by representatives of the typical Taiga florogenetical complex. They populate steep slopes of northern exposure. The above forests can only be found in Tusheti.

Aside from these types of pine forests, the following communities may be mentioned: Pinetum siccum, Pinetum cytisosum, Pinetum calamagrostidosum, Pinetum prasinosum, etc. Oak-pine forests are attributed to a special group of forests; they are confined to lower parts of the pine forest belt (in East Georgia they occur at 800–1,100 m.a.s.l., whereas in Ajara they descend to 300 m).

6.6 Oak Forests

From the coenotical and floristical points of view, the oak forests are among the richest forests of Georgia. According to Dolukhanov (2010), in prehistoric times, the area covered by oak trees was twice as large as it is today. These forests have also suffered from agricultural activities of man.

Oak forests which escaped destruction, are characterized by unsatisfactory regeneration and usually developed on poor soils.

In Georgia, the most abundant oak forests are those dominated by *Quercus iberica*, which is closely related to the widespread European *Q. petraea*. The distribution area of *Q. macranthera* is more limited. Broad-leaved forests of Colchis are characterized by the admixture of *Q. hartwissiana*. Alluvial plains of the valley of river Mtkvari (=Kura) are covered by *Q. pedunculiflora* forests; this species is very close to *Q. robur*, in the lowlands of West Georgia this species becomes replaced by *Q. imeretina*. Remnants of *Q. dschorochensis* communities, which were widely distributed in the past, have survived in Ajara. The characteristic species of crook-stem forest is *Q. pontica* – an ancient relic of the Colchic flora. This species can be found only within the creeping crook-stem forests.

6.6.1 Forests of Georgian Oak (*Q. iberica*)

The distribution area of these forests involves almost the entire territory of Transcaucasia, Daghestan and partly Chechnya-Ingushetia.

As it was pointed out, the south-Transcaucasian species *Q. iberica* displays close taxonomical connections with *Q. petraea*, though they are separated ecologically and phytosociologically.

Q. iberica is typical for eastern and south-eastern semi-arid parts of Transcaucasia; although, it can be encountered also in the Black Sea coastal area of Georgia. Even extreme conditions provided in the areas where the above oak forests come into contact with steppe and mountain-xerophytic vegetation, do not hinder their development. *Q. iberica* forests extend from sea level up to 1,500 and even 1,800 m (in Svaneti).

Being very sensitive to excessive soil moisture and bad drainage, Georgian oak avoids alluvial plains.

Due to the fact that Georgian oak requires much light, in the areas favourable for the development of beech-woods, populations of oak are replaced by more shade-enduring ones of beech.

Unlike in Europe, in Georgia *Fagus orientalis* and *Q. iberica* are never found to grow together.

Of the arboreal components of *Q. iberica* forests the following should be mentioned: *Carpinus caucasica, C. orientalis, Acer laetum, Sorbus torminalis, Zelkova carpinifolia.*

Depending on the degree of anthropogenic succession, derivatives of oak forests are illustrated by *Carpinus caucasica* forests and shibliak vegetation consisting of *Carpinus orientalis, Cornus mas, Cotinus coggygria, Swida (Cornus) australis, Crataegus pentagyna, Spiraea hypericifolia* and *Paliurus spina-christi*. At the last stages, shibliak is substituted by *Bothriochloa ischaemum*. Very often, meadows occupy the areas once covered with oak forests.

Taking into account that *Q. iberica* forests were replaced in many areas by beech-woods, dark coniferous forests, etc., and that a considerable part of the above *Quercus* forests was destroyed by man, many associations of these forests have disappeared by now. Therefore, we assume that oak forests were well developed on the foothills and in the low mountain zone where competition between oak and beech, spruce, chestnut trees, etc. was impossible (Dolukhanov 2010).

Oak forests are very rich floristically, owing to the illumination under the canopy and the heterogeneity of the physico-geographical conditions within the distribution area of Georgian oak. *Q. iberica* forests are confined to rocky sites which promote enrichment of their floristical composition. In limestone regions many local endemics penetrate into the undergrowth of these forests (Dolukhanov 2010).

Dolukhanov specifies the following variants (sub-formations) of *Q. iberica* forest communities:

1. Monodominant forests,
2. *Carpinus orientalis-Quercus iberica* forests,
3. *Carpinus caucasica-Quercus iberica* forests.

Today *Carpinus orientalis-Quercus iberica* forests are confined to the lower parts of the forest belt and occupy the largest area.

Carpinus caucasica-Quercus iberica communities occur on fertile and moist soils at relatively high altitudes. Monodominant forest communities are represented by forests with well-developed undergrowth as well as by those with no undergrowth. The same author specifies by convention the following ecologically different series within monodominant oak forests:

1. Hemi-xerophytic,
2. Xero-mesophytic.

Hemi-xerophytic oak forests develop under the driest conditions, regarding forest vegetation. In prehistoric times, these forests were widely distributed in semi-arid districts of East Georgia. Due to the impact of anthropogenic factors, these forests have always been subject to destruction.

As to the forest maintenance and timbering, regeneration in the above-mentioned forests is unsatisfactory.

Different associations of oak forests include Quercetum iberici multifruticosum siccum and the communities of oak forests with *Cotinus coggygria*.

It is suggested that in the past, oak forests included the following species as undergrowth: *Spiraea hypericifolia, Pyracantha coccinea, Juniperus oblonga.* Georgian oak forests with *Sesleria anatolica* and with *Psoralea bituminosa* are widespread in Abkhazeti.

It should be pointed out, that Iberica-Quercetum genistosum can be observed in east Transcaucasia (with *Genista transcaucasica*) as well as in Northern Colchis (*G. kolakowski, G. abchasica*).

Xerophytic *Q. iberica* forests without woody undergrowth are characterized by presence of rich herbaceous ground vegetation. Associations of these forests have been united by Dolukhanov (2010) into the Iberica-Querceta multiherbosa transcaucasica group. This group involves typical oak forests with forbs, *Q. iberica* communities with various herbs and grasses, and *Q. iberica* forests with various herbs and sedges.

Q. iberica forests with various herbs are confined to river gorges of Pshavis Aragvi, Mtiuletis Aragvi, Gujaretis Tskhali and Nedzviskhevi; these forests extend from 830 up to 1,460 m a.s.l.

Of the arboreal components of these forests the following should be mentioned: *Carpinus orientalis, C. caucasica, Sorbus torminalis, Acer laetum, Picea orientalis, Abies nordmanniana, Chamaecytisus caucasica, Lonicera caucasica,* and *Mespilus germanica*. Composition of the herbaceous undergrowth includes: *Clinopodium vulgare, Veronica peduncularis, Polygonatum glaberrimum, Campanula rapunculoides, Dactylis glomerata,* etc.

The commonest association of *Q. iberica* forests with grasses is the one with various herbs and *Brachypodium sylvaticum*.

Q. iberica forest with various herbs and *Carex buschiorum* is the typical representative of Georgian oak forests with sedges.

The group of associations of oak forests with *Epimedium* chiefly occurs on seaside mountains of Abkhazeti. The presence of *Epimedium colchicum* and *Hypericum xylosteifolium* as constant species is typical for these communities. *Dioscorea caucasica* is a constant species of oak forests restricted to limestone regions of West Georgia. The common species include *Sesleria anatolica, Trachystemon orientale, Ruscus ponticus, Carex transsilvanica, Dorycnium graecum.* Less common species are manifested by *Iris colchica, Dianthus imereticus, Hypochaeris radiata, Psoralea bituminosa, Aristolochia steupii, A. iberica, Primula sibthorpi, Veronica peduncularis, Helleborus abchasicus,* etc. These forests are inhibited by such Colchic shrubs as *Rhododendron luteum, Rh. ponticum, Vaccinium arctostaphylos*. Undergrowth of the forests confined to the cape of Bichvinta (Pitsunda) involves *Erica arborea*. Examples of common lianas are *Lonicera caprifolium* and *Smilax excelsa*. Right at the seashores, *Arbutus andrachne* occurs as an arboreal component of oak forests. The following associations may be regarded as typical for the given group:

Quercetum hypericoso-epimediosum
Quercetum ericosum
Quercetum dioscoreoso-epimediosum
Quercetum multiherbosum-ibericum.

Q. iberica forests with developed woody undergrowth are characterized by the presence of dense thickets of shrubs and sparse undergrowth. The following associations belong to the above-mentioned first type of oak forests: *Querceta azaleosa (Rhododendron luteum), Querceta rhododendrosa (Rh. ponticum), Querceta staphyleosa (Staphylea colchica* is characteristic of West Georgia, whereas *S. pinnata* is of East Georgia). According to Dolukhanov (2010), the second type of these forests involve too many variants to make them subject for any classification.

6.6.2 *Q. iberica* Forests with *Carpinus orientalis*

These communities are widespread in East Georgia. Under the driest conditions they extend from 600 to 1,000 m a.s.l., while in humid districts of West Georgia from 350 up to 700–800 m. At lower altitudes the above communities are restricted to northern slopes, whereas at higher altitudes to southern ones. They can also be found in limestone regions of Abkhazeti and Samegrelo. In East Georgia they occur on different rocks. Dolukhanov (2010) distinguishes three main associations detectable in the oak forests with *C. orientalis*:

1. Carpinuleto-Quercetum xerocaricosum (*Carex buschiorum*),
2. Carpinuleto-Quercetum poosum (*Poa nemoralis*),
3. Carpinuleto-Quercetum brachypodiosum (*Brachypodium sylvaticum*).

The associations of oak forests with *Epimedium* are typical for West Georgia.

In addition the following associations should be mentioned:

Carpinuleto-Querceta ruscosa (*Ruscus colchicus*),
Carpinuleto-Querceta azaleosa (*Rhododendron luteum*),
Carpinuleto-Querceta cotinosa (*Cotinus coggygria*).
Carpinuleto-Querceta sesleriosum (*Sesleria anatolica*) and Carpinuleto-Querceta xerobrachypodiosum (*Brachypodium sylvaticum*) favour the driest climatic conditions.

6.6.3 *Carpinus caucasica-Quercus iberica* Forests

These forests occupy transition zones between oak forests and those of beech. The second stratum is composed of *Carpinus caucasica*, which is a shade-enduring tree. Very little information is available on the typology of the above-mentioned forests because of the poor state of remnants of the climax communities. Dolukhanov (2010) indicates that *Carpinus-Quercus* forests with *Festuca drymeja* are abundant

Table 6.5 Semi-natural *Carpinus-Quercus* deciduous forest near Tbilisi (relevé G-1) (Box et al. 2000) (Location: Just past Kojori along Kojori Road, beyond Botanical Institute (22 May 1999))

T1	10 m	70 %	1,600 m, Slope: 2–15 ° to SW	
T2	6 m	10 %		
S	3 m	20 %		
H	0.4 m	70 %		
			20 × 20 m	
T1	4.4	*Carpinus caucasica*	2.2	*Quercus iberica*
	1.1	*Fagus orientalis*		
T2	2.2	*Carpinus caucasica*		
S	2.2	*Carpinus caucasica*	1.2	*Crataegus monogyna*
	1.2	*Corylus avellana*	1.1	*Acer campestre*
	1.2	*Lonicera caucasica*	+	*Cornus mas*
	+	*Viburnum lantana*	+	*Rosa canina*
	+	*Pyrus caucasica*	+	*Prunus divaricata*
	+	*Malus orientalis*		
H	3.3	*Anthriscus nemorosa*	2.2	*Primula woronowii*
	1.1	*Galanthus caucasicum*	1.2	*Poa nemoralis*
	1.2	*Carex sylvatica*	1.1	*Galium verum*
	+.2	*Polygonatum verticillatum*	+.2	*Lathyrus* sp.
	+.2	*Taraxacum officinale*	+.2	*Cyclamen vernum*
	+.2	*Cruciata laevipes*	+.2	*Lonicera caucasica*
	+	*Quercus iberica*	+.2	*Viola odorata*
	+	*Carpinus caucasica*	+	*Fragaria viridis*
	+	*Acer laetum*	+	*Campanula rapunculoides*
	+	*Acer campestre*	+	*Calamagrostis arundinacea*
	+	*Geum urbanum*	+	*Primula macrocalyx*
	+	*Paris quadrifolia*	+	*Arum albispathum*
			+	*Potentilla* sp.

in East Georgia. Examples of typical species of these forests are *Cornus mas, Corylus avellana, Thelycrania (Cornus) australis.*

Carpinus caucasica-Quercus iberica forests with *Rhododendron luteum* are common in West Georgia.

Most of the natural vegetation has been greatly disturbed if not completely destroyed. An idea of the potential composition of natural *Quercus-Carpinus* forest on slopes around Tbilisi is given by the relevé in Table 6.5, from a secondary forest at about 1,600 m in the hills southwest of Tbilisi. The forest is short (10 m) and young but contained 39 species in the 20 × 20 m plot, including *Fagus orientalis, Acer campestre*, and various forbs characteristic of more mature mesic *Quercus iberica-Carpinus* forests of the region (Box et al. 2000)

The pote ntial composition of subhumid woodland is suggested by the relevé in Table 6.6, from the second terrace above the Mtkvari river about 15 km northwest of Tbilisi. The sample (15 × 15 m) is from a concavity on a lower slope, ranging

Table 6.6 Degraded *Fraxinus-Juniperus* stand in concavity on riverine terrace near Tbilisi (relevé G-12) (Box et al. 2000) (Location: Northwest of Tbilisi, 2nd terrace above river (27 May 1999))

T	7 m	85 %	540 m, Slope: 3–20 ° to SSE	
S	3 m	20 %		
H	0.4 m	60 %		
			15 × 15 m (patch) KF, GN, *EB*, RJL	
T	4.4	*Fraxinus excelsior*	2.2	*Juniperus oblonga*
	1.1	*Quercus iberica*	+.2	*Prunus divaricata*
	+	*Cornus mas*	+.2	*Cotoneaster racemiflorus*
			+	*Prunus spinosa*
S	1.2	*Cotinus coggygria*		
	1.2	*Juniperus rufescens*	1.1	*Prunus divaricata*
	+.2	*Cornus mas*	+.2	*Juniperus oblonga*
	+	*Spiraea hypericifolia*	+.2	*Cotoneaster racemiflorus*
			+	*Prunus spinosa*
H	3.3	*Ruscus ponticus*	2.2	*Brachypodium sylvaticum*
	1.2	*Fraxinus excelsior*	1.3	*Juniperus rufescens*
	1.2	*Campanula rapunculoides*	1.2	*Dactylis glomerata*
	1.1	*Tanacetum vulgare*	1.1	*Fragaria viridis*
	1.2	*Viola odorata*	1.2	*Convolvulus lineatus*
	+.2	*Cruciata laevipes*	+.2	*Draba* sp.
	+	*Astragalus offinalis*	+.2	*Stellaria media*
	+	*Dictamnus caucasicus*	+	*Tragopogon graminifolius*
	+	*Polygonatum verticillatum*	+	*Euonymus caucasicus*
	+	*Thalictrum foetidum*	+	*Astrodaucus orientalis*
	+	*Poa bulbosa* var. *vivipara*	+	*Chelidonium majus*
	+	*Trifolium arvense*	+	*Taraxacum officinale*
	+.2	*Viburnum lantana*	+	*Geranium pallens*
	+.2	*Carex* sp.	+	*Euphorbia glareosa*
	+	*Ajuga genevensis*	+	*Vicia iberica*
			+	*Avena barbata*

from 20 ° above to about 3 ° at the base and sloping toward the south-southeast. The canopy is low (7 m), and the dense cover by *Fraxinus excelsior* is probably not representative of most of the slopes. Otherwise the sample shows a composition which may be fairly typical, including *Juniperus, Prunus, Cotinus* and other shrub genera which were also seen in more natural stands of this vegetation type in eastern Georgia. At 48, the number of species in this plot suggests the diversity of these open woodland landscapes (Box et al. 2000).

Much of the Lagodekhi reserve is covered by deciduous *Quercus iberica* forest, always with *Carpinus caucasica* and usually with *C. orientalis* as well. A relatively rich example of this forest, including also *Fraxinus*, *Acer*, and *Tilia*, is shown in

Table 6.7 *Quercus iberica* forest in Kakhetia, Eastern Georgia (relevé G-13) (Box et al. 2000) (Location: Lagodekhi State Nature Reserve, *Quercus* area (30 May 1999))

T1	25 m	85 %	520 m, Slope: 30 ° to SSE	
T2	8 m	20 %		
S	4 m	20 %		
H	0.4 m	35 %		
			40 × 40 m	
T1	3.3	*Quercus iberica*	3.3	*Carpinus caucasica*
	2.2	*Fraxinus excelsior*	2.2	*Tilia begoniifolia*
	2.2	*Acer laetum*		
T2	2.2	*Carpinus orientalis*	1.1	*Crataegus caucasica*
	1.1	*Cornus mas*	+	*Quercus iberica*
	+	*Fraxinus excelsior*	+	*Cerasus sylvestris*
Vine: +.2		*Hedera pastuchowii*		
Epiphyte: +.2		*Viscum album*		
S	2.3	*Rubus caucasicus*	2.2	*Cornus mas*
	1.2	*Acer laetum*	1.1	*Carpinus orientalis*
	1.1	*Crataegus caucasica*	+	*Mespilus germanica*
	+	*Lonicera caprifolium*		
Vine: 1.1		*Hedera pastuchowii*		
Epiphyte: +		*Viscum album*		
	3.3	*Festuca drymeja*	2.2	*Hedera pastuchowii*
	1.3	*Rubus caucasicus*	1.2	*Laser trilobum*
	1.2	*Aristolochia iberica*	1.2	*Geranium robertianum*
	1.1	*Galium aparine*	1.1	*Carex sylvatica*
	1.1	*Primula woronowii*	1.1	*Asplenium*
	1.1	*Viola odorata*	+.2	*Platanthera chlorantha*
	+	*Carpinus caucasica*	+	*Acer laetum*
	+	*Vicia crocea*	+	*Polygonatum verticillatum*
	+	*Alliaria officinalis*	+	*Dentaria quinquefolia*
			+	*Ajuga genevensis*
	+	*Poa nemoralis*	+	*Lamium album*
	+	*Salvia glutinosa*	+	*Asplenium trichomanes*
	+	*Concolvulus* sp.		
	+	*Scrophularia nodosa*		
	+	*Vicia iberica*	+.2	*Vicia iberica*
	(+)	*Silene wallichiana*		
	(+)	*Dactylis glomerata*	(+.2)	

Table 6.7, from a steep SSE-facing slope at 520 m elevation. In this forest, *Cornus mas* grew to small-tree size, and there was a fairly rich understorey of *Rubus caucasicus, Festuca drymeja, Hedera pastuchowii* vines, and a good variety of herbs. A richer floodplain forest from the same area, with less *Quercus*, is shown in Table 6.8 (Box et al. 2000).

Table 6.8 Floodplain *Carpinus-Acer-Fraxinus-Tilia* forest in Eastern Georgia (relevé G-14) (Box et al. 2000) (Location: Lagodekhi State Nature Reserve, plot "Matsimis Ubani" (30 May 1999))

T1	30 m	50 %	430 m Level (floodplain)	
T2	12 m	50 %		
S	4 m	20 %		
H	0.5 m	60 %		
M		15 %		
			40 × 20 m *EB*	
T1	3.3	*Carpinus caucasica*	3.3	*Acer campestre*
	2.2	*Fraxinus excelsior*	2.2	*Tilia begoniifolia*
	1.1	*Pterocarya pterocarpa*		
Vines: 1.1		*Hedera pastuchowii*	1.1	*Smilax excelsa*
T2	3.3	*Carpinus orientalis*	2.2	*Acer campestre*
	+	*Alnus barbata*		
Vines: 1.2		*Hedera pastuchowii*	+.2	*Smilax excelsa*
Epiphyte: 1.2		*Viscul album*		
S	2.2	*Carpinus orientalis*	1.1	*Carpinus caucasica*
	+	*Fagus orientalis*	+	*Pterocarya pterocarpa*
	+	*Ulmus elliptica*	+	*Sorbus graeca*
	+	*Mespilus germanica*	+	*Sambucus nigra*
	+	*Corylus avellana*	+	*Pyrus caucasica*
	+	*Cornus mas*	+	*Smilax excelsa*
	+	*Calystegia sepium*		
H	2.2	*Asperula odorata*	2.2	*Pachyphragma macrophyllum*
	2.2	*Sanicula europaea*	2.2	*Hedera pastuchowii*
	2.2	*Carex sylvatica*	1.3	*Oplismenus undulatifolius*
	1.2	*Geranium robertianum*	1.2	*Dryopteris filix-mas*
	1.1	*Primula woronowii*	1.1	*Deschampsia cespitosa*
	1.1	*Geum urbanum*	1.1	*Alliaria officinalis*
	1.1	*Galium aparine*	1.2	*Viola odorata*
	+.2	*Asplenium trichomanes*	+.2	*Geranium sylvaticum*
	+.2	*Asplenium pseudolanceolatum*	+.2	*Phyllitis scolopendrium*
	+.2	*Sedum stoloniferum*	+.2	*Convolvulus*
	+	*Acer campestre*	+	*Fraxinus excelsior*
	+	*Carpinus orientalis*	+	*Cormus mas*
	+	*Ajuga orientalis*	+	*Laser trilobum*
	+	*Stellaria media*	+	*Impatiens noli-tangere*
	+	*Euphorbia macroceras*	+	*Lamium album*
	+	*Fragaria vesca*	+	*Orobanche* sp.
	+	*Vicia* sp.		
M	2.2	*Mnium* spp. (3 spp.)		

Table 6.9 Primary deciduous forest with Hyrcanian elements, in Eastern Georgia (relevé G-15) (Box et al. 2000) (Location: Lagodekhi State Nature Reserve, "Rachis Ubani" section (30 May 1999))

T1	28 m	75 %	480 m, Slope: 2–3 ° to WSW	
T2	12 m	10 %		
S	5 m	20 %		
H	0.4 m	90 %		
			30 × 30 m	
T1	3.3	*Carpinus caucasica*	3.3	*Fraxinus excelsior*
	2.2	*Acer laetum*	2.2	*Acer platanoides*
	2.2	*Juglans regia*	1.1	*Acer campestre*
	1.1	*Fagus orientalis*	1.1	*Acer velutinum*
	1.1	*Cerasus sylvestris*		
T2	2.2	*Carpinus caucasica*	1.1	*Carpinus orientalis*
Vine: +.2		*Hedera pastuchowii*		
S	2.2	*Corylus avellana*	1.1	*Fraxinus excelsior*
	1.1	*Fagus orientalis*	1.1	*Acer campestre*
	1.3	*Rubus caucasicus*	1.1	*Euonymus latifolia*
	+.2	*Mespilus germanica*	+	*Ulmus elliptica*
	+	*Cornus mas*	+	*Prunus divaricata*
	+	*Hedera pastuchowii*		
H	2.3	*Asperula odorata*	2.3	*Stachys sylvatica*
	2.2	*Alliaria officinalis*	2.2	*Aristolochia iberica*
	2.2	*Hedera helix*	1.2	*Oplismenus undulatifolius*
	1.2	*Stellaria media*	1.2	*Viola odorata*
	1.2	*Lamium album*	1.2	*Salvia glutinosa*
	1.2	*Geranium robertianum*	1.1	*Euonymus latifolia*
	1.1	*Carex sylvatica*	1.1	*Galium aureum*
	1.1	*Dentaria quinquefolia*	1.1	*Pachyphragma macrophyllum*
	1.1	*Sanicula europaea*	1.1	*Poa nemoralis*
	1.1	*Dryopteris filix mas*	1.1	*Geum urbanum*
	1.1	*Athyrium distentifolium*	+.2	*Phyllitis scolopendrium*
	+	*Fraxinus excelsior*	+	*Acer campestre*
	+	*Primula woronowii*	+	*Euphorbia macroceras*
	+	*Polystichum*	+	*Moehringia trinervia*
	+.2	*Orobanche* sp.	+	*Chaerophyllum maculatum*

Although not significantly richer in species per relevé, a more impressive forest is the primary mesophytic forest in the Rachis Ubani area of the Lagodekhi reserve. A sample from this forest is shown in Table 6.9, from a very slight WSW-facing lower slope with rocky brown forest soil. This forest, which has apparently never been cut, has a canopy composed of *Fraxinus excelsior*, *Carpinus caucasica*, four *Acer* species, *Fagus orientalis*, *Jyglans regia*, and scattered canopy-size individuals of *Cerasus sylvestris*. These are mostly typical European (or eastern European) species, but some elements of the Hyrcanian (east Caucasian-Caspian Tertiary) flora also appear in this forest, such as *Acer velutinum*, *Pterocarya pterocarpa*, and *Hedera pastuchowii*. The canopy on this plot had only 75 % cover, and the herb layer was correspondingly dense (90 %). On a nearby area the canopy reached 30 m and 90 % cover, with more *Fagus* and one individual of *Tilia begoniifolia* with 0.6 m-high plank buttresses. On this area, the T2 and S layers remained sparse and the herb-layer cover dropped to 50 % (Box et al. 2000).

6.6.4 *Q. macranthera* Forests

These forests are dominated by *Q. macranthera* (*sect. Macranthera*), which is known as "highland oak". The distribution area of these species comprises the eastern parts of the Greater Caucasus (including both northern and southern macroslopes), the Lesser Caucasus and mountains of Talysh. It is also found in West Caucasus (Svaneti). *Q. macranthera* forests are developed in north-eastern Turkey and in northern Iran (Elburs mts.).

These forests are confined to mid-mountain, high-mountain and subalpine zones; they are developed under relatively dry conditions at an altitude between 1,450 and 2,400 m a.s.l. (in some places at 2,600–2,700 m). The most favourable conditions for *Q. macranthera* are provided at an altitude of 1,450–1600 m near the upper limit of its distribution. Due to the aggravation of the process of regeneration, caused by overgrazing and other factors, highland oak started to form open woodlands and even crook-stem forests (Sakhokia 1980; Dolukhanov 2010).

Q. macranthera forests grow on mountain forest brown soils; close to timberline they occur on mountain forest-meadow brown soils and those of meadow-steppe.

Q. macranthera forms both mixed and monodominant forests. Birch-oak open woodlands (*Q. macranthera* is accompanied by *Betula litwinowii*) and those of *Sorbus-Betula-Quercus (Sorbus caucasigena) and Acer-Quercus (Acer trautvetteri)* are developed in subalpine and high-mountain belts. Constant species of these forests are manifested by *Carpinus caucasica* and *Fraxinus excelsior*; less characteristic species are *Fagus orientalis,* species of *Tilia. Spiraea hypericifolia, Lonicera caucasica, Berberis iberica, Cotoneaster racemiflora,* etc. frequently occur as underwood in these communities.

Common oak forests are those with:

1. Forbs (*Polygonatum verticillatum, Silene wallichiana, Chaerophyllum aureum,* etc.);
2. Forbs and ferns (*Dryopteris filix-mas*);
3. Subalpine tall herbaceous vegetation (*Senecio platyphylloides, Cephalaria gigantea, Grossheimia macrocephala, Telekia speciosa*);
4. *Calamagrostis arundinacea, Phleum phleoides, Dactylis glomerata.*

Q. macranthera forests with tall herbaceous vegetation are in contact with the vegetation of timberless volcanic uplands of South Georgia. Contact zones are of dry and cold climate; oak forests developed here bear the following names: "oak forests with sheep's fescue [*Festuca ovina* group]", "steppe oak forests with forbs and grasses", "oak forests with sedges" (Dolukhanov 2010). The herbaceous ground vegetation is exhibited by the following species: *Carex buschiorum, Festuca valesiaca, Achillea biserrata,* etc.

Monodominant forests of *Q. macranthera* occur exclusively on steep southward slopes with poor skeletal soil. On the more fertile and moist soils *Q. macranthera* communities are replaced by hornbeam-oak forests (Dolukhanov 2010).

6.6.5 *Q. pedunculiflora* Forests

These forests occupy river valleys. *Q. pedunculiflora (Q. longipes)*, closely allied to the European species *Q. robur*, is an endemic species of Transcaucasia. It frequently occurs as arboreal component of oligo- and polydominant forests. *Q. pedunculiflora* is often accompanied by *Carpinus caucasica*, and also by *Ulmus minor, Acer campestre, A. platanoides, A. velutinum, Tilia begoniifolia,* and sometimes by *Fagus orientalis*. Under moist conditions, *Pterocarya pterocarpa* appears as admixture to these forests (Fig. 6.12).

In past times the above forests covered vast areas in East Georgia, whereas today only few remnants of these, once luxuriant forests, can be found (Dolukhanov 2010). Main massifs of these forests are situated in the valley of the river Alazani (Kakheti).

The following lianas are typical for *Q. pedunculiflora* forests: *Smilax excelsa, Periploca graeca, Clematis vitalba, Vitis vinifera* and *Hedera helix.*

6.6.6 *Q. imeretina* Forests

Quercus imeretina, endemic species of West Georgia, is related to *Q. pedunculiflora*. Distribution area of *Q. imeretina*, which once was well developed in West Georgia and occupied relatively dry areas on the right bank of river Rioni, has reduced under the impact of anthropogenic factors (cutting for timber, destruction of forests for arable land, etc.).

Q. imeretina forms both monodominant and mixed forests with *Carpinus caucasica, Zelkova carpinifolia, Fraxinus excelsior, Pterocarya pterocarpa, Alnus barbata*. Dolukhanov (2010) outlines the following groups of *Q. imeretina* forest types of Georgia:

1. Moist oak forests with *Smilax excelsa,*
2. Fresh oak forests with *Rhododendron luteum,*
3. Dry oak forests with *Carpinus orientalis,*
4. Dry oak forests with *Ruscus colchicus,*
5. Dry oak forests with *Hypericum xylosteifolium,*
6. Dry oak forests with *Zelkova carpinifolia.*

Fig. 6.12 Colchic lowland forest at Paliastomi lake, West Georgia (Photo O. Abdaladze)

6.6.7 *Q. dschorochensis* Forests

Q. dschorochensis forests are restricted to the south-western part of Ajara. Outside Georgia they occur in Chorokhi (Coruh) valley (Turkey). The remnants of these forests have survived only on sunny and dry rocky slopes (Dolukhanov 2010).

6.6.8 *Q. hartwissiana* Forests

Contrary to other Caucasian oaks, *Q. hartwissiana* has never formed monodominant forests. Today it participates in the formation of oligo- and polydominant forests of West Transcaucasia; it frequently occurs as a characteristic component of mixed forests of Colchis. *Q. hartwissiana* grows on rich moist soils. It populates foothills and low-mountain zone within the altitude of 1,200 m. The above-mentioned oak is often accompanied by *Zelkova carpinifolia*.

6.7 Hornbeam Forests

Carpinus caucasica is widely distributed throughout the Caucasus; it prefers fertile soils with good drainage.

These thermophilous species appear as dominants of the second stratum, forming hornbeam-beech and hornbeam-oak forests. *C. caucasica* and *Cerasus sylvestris* (Fig. 6.13a) is a usual component of polydominant forests. It should be underlined, that Caucasian hornbeam is adaptive to various conditions of environment (Shelyag-Sosonko 1980; Box et al. 2000; Dolukhanov 2010).

Fig. 6.13 (**a**) *Cerasus sylvestris*, Lagodekhi National Park, East Georgia (Photo O. Abdaladze), (**b**) *Castanea sativa* with *Rhododendron ponticum* (Photo A. G. Dolukhanov)

Hornbeam forests are very diverse phytosociologically. Dolukhanov (2010) distinguishes the following two classes in hornbeam forests: *C. caucasica* forests with Colchic undergrowth and those without Colchic undergrowth. The first class involves the following communities:

1. *C. caucasica* forests with *Poa angustifolia,*
2. *C. caucasica* forests with *Festuca drymeia,*
3. *C. caucasica* forests with forbs,
4. *C. caucasica* forests with *Trachystemon orientalis.*

Hornbeam forests with Colchic undergrowth are abundant in lower parts of the forest zone, where they are not replaced by beech-woods and dark coniferous forests. The above-mentioned forests include the following associations:

1. *C. caucasica* forests with *Rhododendron luteum,*
2. *C. caucasica* forests with *Vaccinium arctostaphylos,*
3. *C. caucasica* forests with *Rhododendron ponticum,*
4. *C. caucasica* forests with *Laurocerasus officinalis.*

Hornbeam communities with *Rh. luteum* are characteristic of West Georgia, though they can also be met in East Georgia (Kakheti and Aragvi river basin).

C. caucasica forests with *Vaccinium arctostaphylos* occur in West Georgia (Abkhazeti, Upper Svaneti, Guria) between 900 and 1,750 m, whereas those with *Rh. ponticum* are restricted to northern slopes of the lower part of the forest zone of non-limestone regions of West Georgia; they extend from 100 up to 800 m.

Hornbeam forests with *Laurocerasus officinalis* occupy steep stony northern slopes of West Georgia at 300–800 m. They occur on both calcareous and non-calcareous soils.

6.8 Sweet-Chestnut Forests

Castanea sativa forests are developed in both West and East Georgia, but to the west of the country they occupy larger areas. In some localities pure stands of *C. sativa* can be found, but mainly sweet-chestnut occurs as a component of oligodominant beech-sweet-chestnut and hornbeam-beech-sweet-chestnut forests. In prehistoric times, the distribution area of *C. sativa* noticeably exceeded the present one.

Sweet-chestnut forests are chiefly restricted to shady slops. In West Georgia they extend from 100 up to 900–1,000 m a.s.l.; approximately the absolute upper limit of *C. sativa* is situated at 1,400–1,450 m. In East Georgia (Kakhetis Kavkasioni) the lower limit of sweet-chestnut forests is about 400–500 m, and the upper 1,350–1,380 m (Dolukhanov 2010). *C. sativa* forests occur on yellow-brown soils (Urushadze 1987). Though sweet-chestnut generally avoids calcareous soils, it can be found on limestone substratum in several localities in Georgia as well.

The forests of *C. sativa* are made subject to cutting in many regions. Besides, the bad state of *C. sativa* populations can be accounted by the impact of parasitic fungi. Dolukhanov (2010) considers that sweet-chestnut forests of Georgia still need to be adequately studied. He outlines the main association groups of *C. sativa* forests:

1. Sweet-chestnut forests with *Trachystemon orientalis*; these communities found in West Georgia, are represented mostly by beech-sweet-chestnut and hornbeam-beech-sweet-chestnut forests.
2. *C. sativa* forests with ferns (*Dryopteris filix-mas*). These communities occupy stony localities. Here sweet-chestnut is accompanied by *Alnus barbata, Ulmus minor, Acer laetum, Fagus orientalis, Carpinus caucasica.*
3. *C. sativa* forests with forbs (*Sanicula europaea, Galium (Asperula) odoratum, Festuca montana, Paris incompleta, Polygonatum polyanthemum*). These communities are typical for West Georgia. *Hedera helix* is a common liana here.
4. *C. sativa* forests with *Rhododendron luteum*. These communities are widespread in West Georgia. They occupy several localities in mountainous Kakheti.

5. *C. sativa* forests with *Vaccinium arctostaphylos*. These forests are abundant in Abkhazeti. Their arboreal components are represented by *Fagus orientalis* and rarely by *Carpinus caucasica*.
6. *C. sativa* forests with *Rhododendron ponticum*. They are confined to the mountains of West Georgia. Sweet-chestnut is accompanied by beech and especially by hornbeam (*Carpinus caucasica*). *Rh. ponticum* extends up to 3–4 m here (Fig. 6.13b).
7. *C. sativa* forests with *Laurocerasus officinalis*. Contrary to the communities with *Rh. ponticum*, these forests are rarely found. They are characterized by the admixture of *Rh. ponticum* to the dense thickets of *Laurocerasus officinalis*. These communities are confined to West Georgia.

6.9 Forests with *Zelkova carpinifolia*

Zelkova carpinifolia (Fig. 6.14) is a relict of the Tertiary flora. The present-day distribution area of this species involves the refuges of Lenkoran (Azerbaijan) and Colchis; *Z. carpinifolia* sometimes occurs in Kakheti (East Georgia) and Karabakh (Armenia).

Monodominant forests of *Zelkova* are rarely found; usually it forms mixed communities with oak species (*Q. imeretina, Q. iberica, Q. hartwissiana, Carpinus caucasica* and *C. orientalis*).

In many areas in Transcaucasia the upper limit of *Z. carpinifolia* varies in altitudes between 1,200 and 1,700 m a.s.l., whereas in Georgia it lies at 750 m. In West Georgia, *Zelkova* is chiefly confined to plains; very often it is met in forest margins and exposed locations. In Kakheti *Z. carpinifolia* occupies all slopes, excluding northern ones. This species occurs on both thin and rich soils, but avoids saline soils. In West Georgia *Zelkova* forests grow on alluvial-skeleton, sandy and clayey-sandy soils, whereas in Kakheti they are found on both deep brown and thin skeletal soils.

In West Georgia (Colchis) *Zelkova*-hornbeam-oak forests are well preserved. To the east of the country, *Zelkova* forests are represented by *Z. carpinifolia-Carpinus orientalis* communities (Sharashidze 1967)

According to Sharashidze, in West Georgia, *Z. carpinifolia* participates in generating the following associations:

Zelkoveto-Querceta (*Q. imeretina*) hypericosa (*H. inodorum*),
Zelkoveto-Querceta ruscosa (*R. colchicus*),
Zelkoveto-Querceta brachypodiosa (*Brachypodium sylvaticum*),
Zelkoveto-Querceta rhododendrosa (*Rh. luteum*),
Zelkoveto-Querceta juncosa (*J. effusus*),
Zelkoveto-Querceto carpinosa (*C. orientalis*).

The following associations are typical for East Georgia:

Paliureto (*Paliurus spina-christii*)-Zelkoveta astragalosa (*A.brachycarpus*),
Zelkoveto-Carpineta (*C. caucasica*),
Zelkoveto-Crataegeta (*C. pentagyna*) geraniosa (*G. palustre*),
Juglandeto-Zelkoveta clinopodiosa (*Calamintha clinopodium*).

Fig. 6.14 *Zelkova carpinifolia*, Babaneuri Reserve, East Georgia (Photo O. Abdaladze)

6.10 Forests with Yew *(Taxus baccata)*

Taxus baccata was once widely spread in Georgia. Forests with considerable admixture of the yew are preserved in Batsara Reserve (East Georgia). It occupies fresh moist soils, developed on carboniferous rocks. Within the territory of the reservation the yew extends from 900 up to 1,350 m; in some locations it is registered to have reached 1,500 m. In Colchis, the lower limit of *T. baccata* descends to the sea level, whereas in East Georgia the yew never comes down to 700 m. Below the dense canopy, *T. baccata* grows slowly and under the abundant sunlight it is characterized by stag headedness (Dolukhanov 2010) (Fig. 6.15a).

Fig. 6.15 (**a**) *Taxus baccata*, Batsara Reserve, East Georgia (Photo P. Shanshiashvili), (**b**) *Buxus colchica*, Colchis, West Georgia (Photo O. Abdaladze)

6.11 Forests with *Buxus colchica*

Buxus colchica is closely related to *B. sempervirens*, which is widely distributed throughout the southern sector of West Europe and Mediterranean. *Buxus colchica* is confined to West Georgia, especially to the limestone regions of Abkhazeti, Samegrelo, Racha-Lechkhumi. In East Georgia it has been cultivated since pagan times (Fig. 6.15b).

B. colchica extends from sea level up to 1,300 m. It is an arboreal component of scattered forests of hornbeam and other broad-leaved trees. Cochic box, highly valued for timber, undergoes intensive cutting.

The flora of the forests with *B. colchica* which occupy limestone slopes is very rich in local endemics and Mediterranean species. These forests involve such constant species, as *Ruscus ponticus, Hedera helix, Asplenium adiantum-nigrum, Carex divulsa, C. transsilvanica, Cyclamen abchasicum, Veronica peduncularis* and *V. persica* (Dolukhanov 2010).

6.12 Forests with Maples (*Acer velutinum* and *A. laetum*)

Acer velutinum is a typical representative of Hyrcanian (Azerbaijan) forests. In Georgia it occurs only in the valley of river Alazani. The upper limit of *A. velutinum* amounts to 1,000 m. *A. laetum* is confined to West Georgia and extends from sea level up to 1,400–1,600 m (higher 2,400 m it is *replaced by A. trautvetteri*). In East Georgia, *A. laetum* is restricted to moist mountain forests. Both maples occur in beech-woods and mixed forests.

7 Near-Timberline Vegetation

In the Caucasus and, particularly, in Georgia, the upper limit of forests is lowered. Straight-trunk forests of dense stands have definite climatic limits in their vertical distribution. Climate above this upper boundary noticeably varies: summer temperature falls with the increase of precipitation, slopes receive abundant solar radiation with the prevalence of ultraviolet refraction, etc.

Low forests composed of the species of *Abies, Pinus, Picea, Fagus* occur on relatively dry and sunny slopes above the forest belt, whereas subalpine crook-stem forests (*Betula litwinowii, Sorbus caucasigena*) are developed under moist conditions. Low forests have become very thin, due to the fatal effect of even slight anthropogenic influence. Continuous low forests, which have probably existed before, are unavailable in present-day Georgia.

Treeline in this area occurs at about 2,300 m, where the treeline krummholz is composed mainly of *Salix kazbekensis* and *Betula litwinowii*, within a matrix of dense *Rhododendron caucasicum* patches and grassy areas, as suggested by the relevé in Table 7.1. There is also a significant moss cover.

Main forest communities of Georgia which reached the subalpine belt, involve the forests of spruce, pine, beech, fir and oak (*Q. macranthera*).

As it was pointed out, dark coniferous forests extend to the altitude of 2,100–2,200 m on prominent slopes. According to Dolukhanov (2010), the upper limit of pine forests does not surpass the level of the 11 °C – isotherms of the warmest month, and that of low forests: 10.5°C. Under moist climatic conditions, the above-mentioned forests are rarely found at altitudes higher than 2,150 m, whereas in continental areas (Tusheti) they reach 2,400 m a.s.l.

Crook-stem forests of *Fagus orientalis* (Fig. 7.1a) extend to 2,300 m in some areas; at the same time vertical limits of these communities are located at 2,350–2,570 m in the moist mountains of South Colchis. It should be noted that beech does not occur under more continental conditions above 2,200 m, and it never surpasses the level of 11 °C isotherms of the warmest month. Vertical distribution of *Quercus macranthera* attract particular interest. Under dry climatic conditions *Q. macranthera* occupies higher altitudes, than in areas with moist climate. The upper limit of its vertical distribution varies between 2,100 and 2,350 m.

G. Nakhutsrishvili, *The Vegetation of Georgia (South Caucasus)*, Geobotany Studies, DOI 10.1007/978-3-642-29915-5_7, © Springer-Verlag Berlin Heidelberg 2013

Table 7.1 *Rhododendron-Salix* treeline krummholz, above Bakuriani (relevé G-6) (Box et al. 2000) (Location: Treeline above Bakuriani, embankment along road (25 May 1999))

S1	2.0 m	40 %	2,300 m, Slope: 30 ° to NW	
S2	1.0 m	90 %		
H	0.1 m	30 %		
M		40 %		
			5 × 10 m	KF, GN, *EB*, RJL
S1:	3.3	*Salix kazbekensis*	2.3	*Betula litwinowii*
	+	*Salix apoda*		
S2:	5.4	*Rhododendron caucasicum*	1.2	*Vaccinium myrtillus*
H:	2.3	*Deschampsia cespitosa*	2.3	*Calamagrostis arundinacea*
	1.2	*Poa longifolia*	1.2	*Agrostis planifolia*
	1.2	*Alchemilla retinervis*	1.1	*Betonica macrantha*
	1.1	*Polygonum carneum*	1.1	*Athyrium filix-femina*
	+	*Daphne glomerata*	+	*Vaccinium vitis-idaea*
	+	*Oxalis acetosella*	+	*Primula ruprechtii*
M:	3.3	*Hylocomium splendens*	1.2	*Pleurozium schreberi*
	+.2	*Dicranum elongatum*		
lichen	+.2	*Cetraria islandica*		

Betula litwinowii, closely related to *B. pubescens,* is one of the most characteristic and common species of subalpine crook-stem forests of the Caucasus. Its upper distribution limit does not exceed the level of the 11 °C isotherms of August, while in the areas where it is unprotected by snow coat, the 9.5 °C isotherms. The common associate of the above-mentioned species is *Sorbus caucasigena.*

In Bakuriani (the Lesser Caucasus) and Kazbegi (the Greater Caucasus) birch forest vegetation was studied in the subalpine zone (2,200–2,400 m) (Nakhutsrishvili et al. 2009, an unpublished list 12). The following phytocoenoses were distinguished:

Bakuriani 2,200–2,350 m:

Betula litwinowii-Salix kazbekensis
B. litwinowii-Rhododendron caucasicum
B. litwinowii-Sorbus caucasigena
B litwinowii-Acer trautvetteri
B. litwinowii-Salix caprea
B. litwinowii-Tall herbaceous (*Geranium sylvaticum, Anthriscus nemorosa, Astrantia maxima, Valeriana tiliifolia, Heracleum sosnowskyi*, etc.)

Kazbegi 2,200–2,400 m:

B. litwinowii-Salix kazbekensis
B. litwinowii-Salix caprea
B. litwinowii-Sorbus caucasigena
B. litwinowii-Rhododendron caucasicum
B. litwinowii-Vaccinium myrtillus

Fig. 7.1 (continued)

B. litwinowii-Tall herbaceous (*Swertia iberica, Cephalaria gigantea, Aconitum nasutum, Senecio caucasigenus, Valeriana tiliifolia*, etc.)

B. litwinowii with meadow elements: *Calamagrostis arundinacea, Deschampsia flexuosa, Alchemilla laeta, Betonica macrantha*, etc.

The Colchic endemics *Betula medwedewii, B. megrelica* and *Quercus pontica* are less frost-resistant, but they are well adapted to cold and moist subalpine summer. As components of crook-stem forests they reach 2,300–2,700 m, whereas under more continental conditions their upper limit is lower. Dolukhanov ascribes this dependence of upper limits on continentality of climate to the specific features of each species (Figs. 7.1b, c and 7.2a–c).

Open woodlands, low forests and crook-stem forests are very diverse and floristically rich. In the subalpine belt these communities are restricted to prominent

Fig. 7.1 (**a**) Crook-stemmed Fagetum (Photo A. G. Dolukhanov), (**b**) *Quercus pontica*, Colchis, West Georgia (Photo D. Kharazishvili), (**c**) Timberline, Kazbegi, the Central Caucasus (Photo U. Bohn)

slopes; they are rarely found at altitudes higher than 2,150 m. Fir and spruce forests lose their productivity with the increase of altitude. Under the impact of anthropogenic factors, pine forests and those of *Quercus macranthera* have almost everywhere and quite irregularly retreated from their natural upper boundaries.

The upper limit of straight-trunk beech forests is situated at about 2,050–2,100 m (rarely at higher altitudes). At altitudes higher than approximately 2,200–2,300 m, beech forests are represented by dwarf semi-creeping formations; trees are pressed to the ground under the snow mass and lower parts of their trunks are rooted. The fact, that due to the increase of altitude, trunks of beech become dwarfed, is worth special interest. Crook-stem forests of *Fagus orientalis* are characteristic of Colchis. Several creeping, vegetative-migratory representatives of undergrowth (*Vaccinium arctostaphylos, Ilex colchica, Laurocerasus officinalis, Ruscus colchica,* and rare *Rhododendron ponticum*) grow under the canopy of these above-mentioned forests.

Fig. 7.2 (continued)

In the areas with less contrasting conditions, crook-stem forests of beech are in contact with *Sorbus-Betula* forests as well as with thickets of such Colchic endemics as *Rhamnus imeretina, Corylus colchica, Sorbus subfusca*, and the species of *Salix* distributed in highlands. Low forests of *F. orientalis*, unavailable today, might have existed before.

Crook-stem forests of *Betula litwinowii* are the most typical communities of subalpine belt. This endemic species of the Caucasus is closely related to *B. pubescens*.

Under relatively continental climatic conditions, the lower limit of *B. litwinowii* comes to approximately 2,000 m a.s.l., and the upper boundary varies between 2,400 and 2,600 m. At the same time, *B. litwinowii* occurs also at lower altitudes; it

c

Fig. 7.2 (**a, b, c**) Crook-stemmed *Betula medwedewii*, Bakhmaro, West Georgia (Photo A. G. Dolukhanov)

comes down by moraines, scree, rocky slopes, etc. Although *B. litwinowii* is a light consumer, it occupies the slopes of northern exposure to secure protection by snow cover in winter.

The specific feature of *B. litwinowii* low forests is that they occur exclusively on dry mountains of East Georgia. Today these communities are almost completely destroyed.

Crook-stem forests of *B. litwinowii* contrary to beech communities are rarely monodominant: the constant species of these communities is *Sorbus caucasigena.* This species is represented in the Caucasus by the following two races (microspecies): *S. caucasigena* and *S. boissieri.* The distribution area of another Caucasian birch, *B. raddeana,* includes Tusheti mountains and the upper parts of the rivers Didi Liakhvi, Aragvi and Tergi river basin. In Georgia, *B. raddeana* extends up to 225 m. In the lower sector of the subalpine belt *B. litwinowii* is found along with *Acer trautvetteri, Salix caprea, Padus avium,* etc. In Colchis *B. litwinowii* is accompanied by *Sorbus subfusca, Rhamnus imeretina, Ribes biebersteinii,* and *Rubus buschii.*

Rhododendron caucasicum, an endemic of the Caucasus, is a characteristic species of crook-stem forests of birch (Figs. 7.1c and 7.3a, b).

It is specific that birch forests with *Rhododendron* are characterized by presence of representatives of the flora of taiga. Examples are *Huperzia selago, Lycopodium annotinum, L. clavatum, Diphasiastrum (Lycopodium) alpinum, Gymnocarpium dryopteris, Cystopteris montana, Listera cordata, Deschampsia flexuosa, Orthilia secunda, Pyrola minor, Vaccinium vitis-idaea.*

Vaccinium myrtillus, V. uliginosum and *Oxalis acetosella* are the exclusive species of these communities.

Fig. 7.3 (**a**) *Rhododendron caucasicum*, Kazbegi, Central Caucasus (Photo O. Abdaladze), (**b**) *Rhododendron caucasicum,* Colchis, West Georgia (Photo A. G. Dolukhanov)

Bryophytes growing in the above-mentioned forests are very diverse and peculiar for coenotic variability. These communities are rather rich in species (over 600) (Dolukhanov 2010).

Kvachakhidze (1979) outlines 15 associations of birch forests situated on the southern slope of the Greater Caucasus. These associations may be united into the following 5 groups, such as:

1. Birch forests with evergreen undergrowth,
2. Birch forests with summergreen undergrowth,
3. Birch forests with tall herbs,
4. Birch forests with forbs,
5. Birch forests with grasses.

Fig. 7.4 *Rhamnus imeretina*, Colchis, West Georgia (Photo Z. Manvelidze)

Fig. 7.5 Crook-stemmed *Quercus pontica,* Colchis, West Georgia (Photo A. G. Dolukhanov)

West Georgia (especially South Colchis) is characterized by subalpine vegetation, which is unavailable in any other areas of West Eurasia (Kolakovsky 1980; Dolukhanov 2010). Only creeping and semi-creeping woody plants grow in this subalpine belt, and among them, there are light-demanding and shade-bearing trees. Light-demanders are exemplified by *Betula medwedewii, B. megrelica, Quercus pontica, Rhamnus imeretina* (Fig. 7.4), *Sorbus subfusca* and *Corylus colchica.*

Crook-stem forests of *Quercus pontica* (Fig. 7.5) and *Betula medwedewii* emerge from under the snow cover only in June. Vegetative reproduction is typical for these species (Dolukhanov 1956; 2010).

The following shade-enduring plants frequently occur as underwood in the creeping crook-stem forests: *Rhododendron ponticum, Laurocerasus officinalis, Ruscus colchicus, Ilex colchica.*

At present, relatives of the species forming crook-stem forests and Colchic undergrowth have mainly survived in the mountains exposed to the summer monsoon in eastern and south-eastern Asia, in the Appalachians of North America and on Maccaronesia. Thus, the Colchic refuge is isolated from the other centres of these floristic elements. This leads to the suggestion that the corresponding endemic species of the subalpine belt of Colchis belong to the relics of a rather remote past (Dolukhanov 1956; 1966a, b; 1980).

Below are given Betuleta relevés from Bakuriani and Kazbegi compiled in 2009–2011 (Tables 7.2 and 7.3).

Interesting example of Betuleta (*Betula medwedewii*) represented in the river Chirukhistskali gorge, Adjara (Colchis) was described by D. Kharazishvili (2005). *B. medwedewii* and forests dominated by this species are quite widespread in the subalpine belt of the seaside part of southern Colchis (Guria, Adjara, Lazistan of Turkey). As reported by Dolukhanov, *B. medwedewii* is a conservative relict of the ancient flora. It is referred to a section *Costatae*, other representatives of which are in the main distributed in southern and south-eastern Asia. A number of its species occur in the Himalayas and mountains of the North America. According to this author, Colchis is the only place throughout the vast areas of the west Eurasia, where two representatives of this section, namely, *B. medwedewii* and *B. megrelica* have remained. The frequency of the latter species is lower and it occurs in northern Colchis. *B. medwedewii* represents the isolated, vegetatively motile, semi-prostrate life form, which is well adapted to highly humid (perhumid) climate.

Fago-Betuleta are elfin forests. The coenoses of such composition, characterized by the elfin form, such distribution pattern and, first of all, not subjected to human intrusions are rare (Table 7.4). These coenoses are formed on very steep slopes, at quite high altitudes. They are primary owing to the high inclination degree of slopes. Woody plants are diverse. The maximum tallness of the trees is 3 m, some trees have the elfin form (beech), others do not exceed shrubs (birch, willow, poplar). *Gypsophila tenuifolia* is a noteworthy endemic (Caucasian).

7.1 Treeline Biotopes

On the Central Greater Caucasus the natural treeline ecotone consists of open and elfin crooked-stemmed (Krummholz) birch forests, tall herbaceous vegetation, scrub and meadows. The treeline vegetation in this part of the Caucasus is strongly degraded and lowered (at the average by 200–400 m) because of long-term overgrazing, tree cutting, etc. (Dolukhanov 1966a, b; 1978; 2010; Nakhutsrishvili 1999). The treeline in the Central Caucasus is characterized by high level of plant diversity and endemism (Kharadze 1948; Sakhokia and Khutsishvili 1975; Gagnidze 2000). In the Kazbegi region, the ecological state of the treeline ecotone can be assessed as normal only on certain massifs. These forests have been

Table 7.2 Betuleta, Bakuriani (Leser Caucasus, Trialeti)

Nr	1	2	3	4	5	6	7	8	9	10	11	12	13	14	15	16
Fall	50			40	40	45	35	45	40				20	70	80	30
Exposure	N	N	N	N	NNE	N	NE	NE	N	N	N	N	NW	N	N	NNE
Plot size (m^2)	50		200	100	50	50	50	100	100		100	100	200	100	100	200
Height tree		8					8		10	15	10	10	10	7	7	7
Tree cover (%)		20					20		70	60	40	70	80	70	80	80
Shrub height	4	4	4	4	4	5	4	3		4						
Shrub cover (%)	80	60	20	80	60	70	30	30		10						
Herb height	1	1	120	60	60	60	1	1	1	150	150	120	1	120	120	120
Herb cover (%)	70	90	70	70	90	90	90	90	90	90	90	80	85	70	80	80
T + S																
Betula litwinowii	5	2/2	1		2	3	1	1	4	3/1	3	4	5	4	5	4
Salix kazbekensis	2	3														
Salix caprea		+	2				2/1	2	+	2	2	+	+	+	+	2
Sorbus caucasigena			1	5	3	1	2	2	+				1		+	2
Salix kusnetzowii														+		
Ribes biebersteinii			+	+		+	+							+	+	+
Acer trautvetteri							+		0/1		2					
Quercus macranthera													+			
Fagus orientalis														+		
H																
Rubus saxatilis	+															
Ranunculus caucasicus	+	1	1				+	+		+		+	+			
Polygonum carneum	+	+					+			+			+		+	+
Calamagrostis arundinacea							2	1	1			1			1	
Swertia iberica	1	2														
Deschampsia flexuosa									+							

Chaerophyllum roseum	†															
Sorbus caucasigena												†			+	
Sedum stoloniferum	†	2														
Alchemilla oxysepala	1	1					†	1			2		†	1	2	
Hieracium pannoniciforme								†					†			
Veratrum album				†		†	2		1	2	†	†	†	†	†	1
Myosotis arvensis		1	+				+						+		+	
Cruciata glabra	2												+			
Cephalaria gigantea	2	2					+			+	+	1	2	+	1	+
Vicia grossheimii	+															
Fritillaria lutea											+					
Vaccinium myrtillus			+								1					
Trollius ranunculinus			+			+	1	+	1	2						
Betonica macrantha	3	3						+	+	+			+			+
Veronica gentianoides	+	+														
Macrotomia echioides	+	+														
Primula amoena			1								+					
Pimpinella rhodantha											+		+			
Cirsium frickii						+										
Valeriana alpestris		1														
Primula macrocalyx			1													
Scabiosa caucasica																+
Festuca varia subsp. *woronowii*	2	2										1			1	
Dolichorrhiza caucasica	+															
Rhododendron caucasicum			3	2	5	2	2	5								
Geranium sylvaticum						+	2	1	1			2		+	+	+
Poa iberica				+								+			+	

(continued)

Table 7.2 (continued)

Nr	1	2	3	4	5	6	7	8	9	10	11	12	13	14	15	16
Lathyrus cyaneus	+															
Bromopsis variegata											+					
Alchemilla sp.			+	+		1										
Polygonatum verticillatum									1							+
Lapsana grandiflora			+			1	1	+	1	+	+	+	+	1	1	1
Aconitum nasutum				+		1	+		1		+		+	+		+
Vicia balansae										+		2	1	1	1	1
Festuca sylvatica		2							1		2		2		2	3
Senecio caucasigenus							+			1						
Anthriscus nemorosa							+		+				3			
Chaerophyllum maculatum						+							2	1	1	1
Cirsium obvallatum	+	+					+			+	+		+	+		
Silene vulgaris			+				+		+	+	+		+	+	1	
Vicia sepium									2	1			+	+		
Urtica dioica				1							+					
Pedicularis condensata								1	+	+		+	+	+		
Salix caprea											1					
Trifolium ambiguum								+			+					
Inula orientalis	+															
Poa nemoralis	+										2		1		+	+
Dryopteris filix-mas				2					+					1		
Pyrethrum roseum	+											+				
Ranunculus oreophilus															+	
Trifolim trichocephalum												1		+		
Angelica tatianae			+								+			+	1	+
Geranium ibericum													1			

Lotus caucasicus								+			+					
Campanula collina								+								
Ligusticum alatum											1					
Leontodon hispidus											+					
Valeriana tiliifolia	2			1	3		2		2			2		3	3	1
Heracleum asperum			+						+	1	+	2	1	1		2
Cirsium vulgare	+	+							+							
Cerastium hemschinicum	+															
Vicia crocea		+														
Epilobium angustifolium			2								+					
Milium effusum			2	3		2	2		2	2		2		1		
Rubus idaeus			1				1									
Rumex alpinus			1	+		+	+				1			+	1	
Doronicum macrophyllum			+					+		1						
Senecio othonnae			+			2										
Cardamine uliginosa			+	+		+										
Oxalis acetosella				2		+										
Rhynchocorys elephas				+												
Epilobium alpestre				1		+										
Geum Cf. *montanum*				+												
Symphytum asperum				+						1	1					+
Heracleum wilhelmsii						2					+				+	
Cruciata laevipes						+					+					
Cicerbita macrophylla						+										1
Doronicum macrophyllum						1						+				
Inula grandiflora							2	1								
Saxifraga cymbalaria							+	+	+					1	1	
Aquilegia caucasica							+									
Erigeron alpinus								+								

(continued)

Table 7.2 (continued)

Nr	1	2	3	4	5	6	7	8	9	10	11	12	13	14	15	16
Gentiana schistocalyx								+								
Doronicum orientale								+		+						
Traunsteinera globosa								+								
Astrantia maxima								1		1			2	+		
Petasites albus									2			1				
Galium odoratum									+				+	+		
Primula amoena									+							
Athyrium filix-femina									1							
Veronica gentianoides.									+							
Dactylorhiza maculata									+							
Veronica chamaedrys									+							
Galega orientalis										2	2					
Rhynchocorys elephas										+						
Cerastium sp.										1	+	+			+	+
Dactylis glomerata										+			+		+	+
Heracleum sosnowskyi													+	2		1
Grossheimia macrocephala										+						
Knautia montana											1					
Lamium album											+					
Bunias orientalis											2					
Cicerbita macroyphylla											+					
Achillea millefolium											+					
Trifolium medium											+		+			
Trifolium pratense											1					
Betula litwinowii											+			+		
Inula helenium											+					
Campanula latifolia											+	+		2	1	+

Tripleurospermum caucasicum		+					
Phleum pratense		+					
Tragopogon graminifolius		+					
Linum hypericifolium	+	+		+			
Trisetum flavescens				+			
Bupleurum nordmanianum				+			
Acer trautvetteri				+		+	+
Seseli transcaucasicum				+			
Senecio rhombifolius					+		
Aconitum orientale					+		
Ptarmica ptarmicifolia					+		
Barbarea minor					+		
Heracleum wilhelmsii					+		
Veronica crista-galli					+		
Veronica longifolia					+		
Trifolium repens						+	
Galium aparine						+	
Poa alpina						+	
Rumex arifolius						+	
Poa longifolia						+	
Rosa sp.							1
Hesperis matronalis							+
Lilium szovitsianum							+
Euphorbia macroceras							+
Aconitum orientale							+
Hypericum hirsutum							+
Lathyrus aureus							2

Table 7.3 Betuleta, Kazbegi (Central Greater Caucasus)

Nr	**1**	**2**	**3**	**4**	**5**	**6**	**7**	**8**	**9**	**10**	**11**	**12**	**13**	14	15	16	17	18	19	20	21
Fall	15	20	25	30	25	25	20	25	25	30	35	40	35	45	25	40	60	40	55	20	30
Exposure	N	N	NW	NW	N	NW	N	N	N	NE	N	N	N	NE	NE	NE	NE	N	N	N	N
Plot size (m^2)	100	100	25	50	50	100	100	100	100	100	100	100	100	100	50	50	50	100	100	100	100
Height tree					15	15	12	12	15	15		6						10	10	10	
Tree cover (%)					20	75	70	60	75	70		40						70	50	50	
Height shrub	4	4	3	2	4	3	2	2	2	2	5	2	5	150	150	150	150	150		3	5
Shrub cover (%)	80	80	90	60	80	20	20	15	30	10	25	10	40	60	60	70	90	5		20	80
Height herb	40	40	50	40	1	1	50	1	50	50	1	1	1	120	70	70	70	120	120	120	120
Herb cover (%)	60	50	70	100	80	80	80	90	80	80	90	80	90	100	100	100	100	90	80	70	70
T + S																					
Betula litwinowii	5	5	5	3	3	5/1	4	4/2	4/3	4/2	2	3	3	1	1	1	+	3	3	2	3
Salix kazbekensis	2	1	2	2			+								+		1				
Salix caprea					3				+		1	+						2			1
Sorbus caucasigena					+	2	1	1	+	+		1	+							1	
Rosa mollis					+	+	2		+		1										
Salix kusnetzowii					2									2	1						
Lonicera caucasica						+															
Populus tremula									1												
Rhododendron caucasicum															2	2	3			2	2
Ribes biebersteinii																				+	
Rosa oxyodon																		1		+	
Rubus idaeus																1					
Vaccinium myrtillus														1		2	2				
H																					
Rubus saxatilis	2	2	2	+	1	2	3	3	2		2		+	1				+	+	+	+
Ranunculus caucasicus	+		+	1	+		+	+	+	1	1	1	1	+	+						

Polygonum carneum	1	1		1	1	1	1	+	1	1	1	+		+		+		1	+	+	
Calamagrostis arundinacea	1	1	2				1	3	3	1	2	2	3			1					
Swertia iberica	+		+		+	+															
Deschampsia flexuosa	+	+		+			+							2	1						
Chaerophyllum roseum	+	+									2	2	1								
Sorbus caucasigena	+	+												+							
Daphne glomerata	1	1	+											1	+	1					2
Sedum stoloniferum	+					+	+										+	+	+		
Alchemilla laeta	1	1	+	1			3	2	2	3			2					1			
Hieracium pannoniciforme	1	+		+		1	1	1	1	2	1	1	1	+		1					
Veratrum album	+	+			3	+	1	+						+				1	+	1	
Salix kazbekensis	2																				
Anemonastrum fasciculatum	1													1		1			1		
Myosotis arvensis	+	+		+	+		+			+	+							+	+	+	
Cruciata glabra	+	+	+				+	+	1	2	+	1	+								
Cephalaria gigantea	+			1			+							+							
Cerastium hemschinicum	+																				
Vicia grossheimii	+	+	1	+			1	1	1			1									
Fritillaria lutea	+	+																			
Vaccinium myrtillus	+	1	1	3			+	+													
Viola odorata	+		+																		
Trollius ranunculinus	1						1			+											
Betonica macrantha	+	1	+			+	2	2	2	+		+	+	1		1		+	1	+	+
Veronica gentianoides	r			+		+	+				+										
Macrotomia echioides	+		+																		
Dactylorhiza amblyoloba	+									+											
Primula amoena		1		+																	

(continued)

Table 7.3 (continued)

Nr	**1**	**2**	**3**	**4**	**5**	**6**	**7**	**8**	**9**	**10**	**11**	**12**	**13**	14	15	16	17	18	19	20	21
Cerastium holosteum		+																			
Pimpinella rhodantha		+	+		+			+	+	1											+
Carex leporina		+																			
Vaccinium vitis-idaea		+													2	+	1				2
Valeriana alpestris		+		+												+	+		1	1	
Primula macrocalyx		+					+			+											
Scabiosa caucasica			+												+						
Festuca varia subsp. *woronowii*			2	2										2		1					
Dolichorrhiza caucasica			+	2	+	1		+													
Rhododendron caucasicum				3																	
Cirsium horridum				1																	
Salix kusnetzowii				1																	
Centaurea cheiranthifolia				+										+							
Geranium sylvaticum				+	+	2	1	1			+			1				1	1	1	
Poa iberica				+																	
Viola somchetica				+																	
Lathyrus(Orobus) cyaneus				+							+			1		+					
Bromopsis variegata				+											2	1					+
Anthoxanthum alpinum				+							1										
Carex meinshauseniana				+												1	1				
Alchemilla oxysepala					+						+	2								+	1
Polygonatum verticillatum					+	+	+	+	+		+	+	2							1	
Lapsana grandiflora					+	+	+		+	+				+							
Rumex acetosa					+							+	+								
Aconitum nasutum						+	+	1										+			
Vicia balansae						3													2	1	

Pyrola media					+		+	+											
Festuca sylvatica					2								+				1	2	2
Lathyrus pratensis					+		+			1	1	1							
Senecio caucasigenus					+							r							
Anthriscus nemorosa						+	+	2	+										
Chaerophyllum maculatum						+		+	+										
Cirsium obvallatum	+				+	+	+	+					1						
Silene vulgaris						+			+	+		+							
Vicia sepium						+	+												
Solidago virgaurea						+	1							+		+			
Urtica dioica						+											1		
Luzula pilosa						+													
Helictotrichon pubescens						+							1		1	+			
Pedicularis condensata						r				+									
Geranium depilatum							+												
Daphne mezereum							+												
Salix caprea							+												
Trifolium ambiguum							+	+	2		2	+	+	+					
Platanthera chlorantha							+	+	+	+									
Inula orinetalis							+		+		+								
Carex pallescens							+												
Dactylorhiza euxina							+	+											
Cicerbita racemosa								+											
Lotus corniculatus								+											
Carum meifolium								+											
Rosa mollis								+											
Poa nemoralis									1		3	2							

(continued)

Table 7.3 (continued)

Nr	**1**	**2**	**3**	**4**	**5**	**6**	**7**	**8**	**9**	**10**	**11**	**12**	**13**	14	15	16	17	18	19	20	21
Dryopteris filix-mas										+											
Pedicularis chroorrhyncha										+											
Trifolium alpestre										1	2	1									
Poa pratensis										2											
Festuca ovina											1				1	+	2				
Koeleria luerssenii											+										
Phleum montanum											1		+								
Pyrethrum roseum											1	2	+								
Ranunculus oreophilus											1	+	+								
Trifolim trichocephalum											+	+	+	+							
Angelica tatianae											+	+						1			
Taraxacum officinale											+										
Geranium ibericum											2	1									
Lotus caucasicus											+										
Pastinaca armena											1			+							
Campanula collina											+	1	+	1		1					+
Ligusticum alatum											+	1	1								
Cerastium arvense											+	+									
Leontodon hispidus											1			+	+	+					
Valeriana tiliifolia												2						1			
Heracleum asperum												+	3					+			
Leontodon danubialis												+									
Heracleum roseum												+									
Vicia alpestris													+								
Agrostis planifolia														2	+	1	+				
Alchemilla retinervis																+					

Alchemilla rigida			1	1				
Alchemilla sericata	1	1						
Allium victorialis							3	
Anemone speciosa		+		+				
Antennaria caucasica		+						
Anthemis sosnovskyana		+						+
Anthoxantun alpinum			2	1				
Anthyllis variegata		1						
Arenaria lychnidea	+			+				
Aster alpinus		+	+					
Bromopsis variegata					1			
Campanula biebersteiniana				+				
Carex tristis		2	1					1
Carum caucasicum		1	1	+				
Chaerophyllum aureum					+			
Cirsium pugnax					+	+	1	
Dolichorrhiza renifolia						1		
Dryas caucasica								1
Empetrum caucasicum		1	1	2				
Galega orientalis					2			
Galium rotundifolium					1			
Gentiana pyrenaica			+	+				+
Gymnadenia conopsea	+	+	1					
Helictotrichon asiaticum	1	2						
Hesperis matronalis					+			
Iris sibirica							+	
Kobresia capilliformis								+

(continued)

Table 7.3 (continued)

Nr	**1**	**2**	**3**	**4**	**5**	**6**	**7**	**8**	**9**	**10**	**11**	**12**	**13**	14	15	16	17	18	19	20	21
Lilium georgicum																			1		
Luzula pseudosudetica																	+				
Minuartia caucasica															+						1
Minuartia oreina																+	+				
Oxalis acetosella														+							
Oxytropis cyanea														1		+					
Pedicularis wilhelmsiana																	+				
Poa alpina														1			1				
Poa longifolia																1					
Polygala alpicola																1					
Polygonatum orientale																			2		
Polygonum viviparum															+						
Primula amoena																			+		
Rhinanthus minor															1	+					
Podospermum alpigenum															+						
Senecio rhombifolius																		1			
Sibbaldia parviflora																	1				
Sibbaldia semiglabra														+							
Silene linearifolia														+							
Traunsteinera sphaerica															+						
Trifolium canescens																					1
Trifolium repens																1					

Table 7.4 Chirukhistkali Canyon, Mt. Tbethi (Kharazishvili 2005)

T 3 m	85 %		2,380 m, 60 ° NE	N 41 °32′640″
S 1.5 m	60 %			
H 0.8 m	90 %			E 42 °11′512″
T	*Betula medwediewii*	4		
	Fagus orientalis	4		
	Picea orientalis	+		
	Salix caucasica	+		
	Populus tremula	+		
m	*Rhododendron luteum*	3		
	Vaccinium arctostaphylos	2		
	V. uliginosum	2		
	Juniperus pigmaea	2		
	Sorbus boissieri	2		
	Rubus hirtus	+		
H	*Polystichum lobatum*	3		
	Gypsophila tenuifolia	2		
	Pyrola minor	2		
	Achillea millefolium	2		
	Alchemilla retinervis	+		
	Dentaria quinquefolia	+		
	Valeriana alliariifolia	+		
	Draba hispida	+		
	Phleum pratense	+		
	Gentiana schistocalyx	+		
	Senecio platyphylloides	+		
	Aruncus vulgare	+		
	Veratrum lobelianum	+		
	Geranium psilostemon	+		

protected because of their religious significance and they are called “Holy Forests”. According to the degree of naturalness of these fragments, they should be attributed to the first level of hemeroby: natural and close to natural (Nakhutsrishvili et al. 2004a, b).

A study was conducted to reveal and characterize the main treeline biotopes of the Kazbegi region (the eastern part of the Central Greater Caucasus).The studies were carried out in the Kazbegi region which is situated on the north-facing major slope of the Main Watershed Range of the Greater Caucasus (N 42°39′; E 44°37′), on the valley of the river Tergi. This region is located on the highest and geomorphologically the most complex central part of the Greater Caucasus. The topography is formed by Jurassic rocks, Palaeozoic and even older granites, young lava and moraines. The mountain massives of the Kazbegi volcanic area are overlayed by Quaternary and Contemporary glacial or river deposits and stone falls as well as strong accumulation of calcareous tuffs and travertine. Glacial and other continental Wьrm deposits occur in many places. The elevation range in the region varies from 1,210 m up to 5,033 m a.s.l. (the highest peak is Mkinvartsveri); the average elevation is 2,850 m a.s.l.. Brown and light brown skeletal soils of medium and shallow depth and degraded forest and secondary meadow soils are predominant in the treeline ecotone.

In the Kazbegi region the natural treeline ecotone is located between 2,450 and 2,550 m a.s.l. The climate between 1,900 and 2,600 m a.s.l. is moderately humid. Winter is relatively dry and cold, and summer is short. The mean temperature of the warmest months (July, August) ranges from 10 °C to 14 °C (the absolute maximum is 33 °C). The mean temperature of the coldest month (January) is −11 °C (the absolute minimum is −32.5 °C). The number of days with freezing temperature is 124 per year. Stable snow cover persists for 5–7 months and reaches its maximum depth (115–120 cm) in March. The average annual precipitation is 1,000–1,200 mm, and the summed precipitation during May-August is about 100 mm. The mean air humidity in summer is 75 %. Mist is frequent in the area (135 misty days per year), especially in summer. Winds of mountain-gorge type prevail. Duration of growing season is 6 months (Nakhutsrishvili 1999, 2003).

Phytosociological surveys have been carried out according to the method of Braun-Blanquet (Braun-Blanquet 1964) during 1961–2004. For biotope classification we were guided by the works of Holzner (Holzner (Hrsg.) 1989), Pott (1996) and Pedrotti (1998). We followed the work of Pott (1996) to individualise the hemeroby levels. Priority habitat types were distinguished according to 92/43 EEC from September 1993. Mean and standard deviation were determined for every data set (Nagy and Grabherr 2009).

The total area of the subalpine zone in the Kazbegi region amounts to 245.75 km^2. About 1,100 species of vascular plants are recorded in the whole region. The species number in the subalpine zone (from 1,800–1,900 up to 2,400–2,500 m a.s.l.) is 595, of which 33.2 % are endemics. The total area covered by forest is 8,707 ha.

Below are given some characteristic features of the main treeline biotopes of the Kazbegi region (the Central Greater Caucasus).

7.2 Biotope Complex: Biotopes of Woody Plants

Birch forest (*Betula litwinowii*). The range of the birch, which belongs to the periglacial plants, was remarkably diminished in the xerothermic period of the postglacial epoch and the species has remained only in the subalpine zone (Dolukhanov 1966a, b; 2010). At present the birch is dispersed at the treeline of the whole Caucasus, although the areas occupied by the species are not large. Individual trees grow at an elevation of 2,550 m in some places and single specimens broken down by avalanches may be found even in the middle-mountain forest belt. ***Habitat:*** Humid north-facing slopes (10–25° of inclination) with stable snow cover. Deep and slightly skeletal brown soils covering volcanic rocks, light, loamy; humus content in the upper 10–20 cm layer is about 7 %. ***Distribution:*** fragmentary (between 1,850 and 2,200 m): the Liphu forest (above the village

Gergeti), the Sno gorge, the surroundings of the village Sioni and village Djuta; the forest is best developed in the Devdoraki and Khde gorges; a fragment is found on northwest-facing slopes in the Gudauri region (2,000–2,300 m). ***Characteristic species:*** *Betula raddeana, Salix caprea, Heracleum roseum, Aconitum nasutum, A. orientale, Swertia iberica, Geranium sylvaticum, Campanula latifolia, Dolichorrhiza caucasica, Senecio propinquus, Aquilegia caucasica, Vicia balansae, Lathyrus roseus, Cephalanthera longifolia, Platanthera chlorantha.* ***Ecological importance:*** Water-regulatory, protection against avalanches and erosion. ***Threats:*** Tree felling, intensive grazing, construction of roads, climate global change (warming) (Akhalkatsi et al. 2006).

Elfin crooked-stemmed birch forest (*Betula litwinowii*). These forests occur at the treeline, particularly, in areas with abundant snow in winter. The elfin form is due to the weight of the snow covering trees in winter. Such a life form helps woody plants to adjust to severe winter conditions. In the period following the Ice-age the range of the birch forests diminished drastically; they have remained only in the subalpine belt. Its upper distribution limit does not exceed the level of the 11 °C isotherms of August, while in the areas where it is unprotected by snow coat, the 9.5 °C isotherms (Nakhutsrishvili 1999). ***Habitat:*** North-facing slopes with deep snow cover as well as various inclinations (15–70°). Peat brown soils of medium depth, which are situated mainly on volcanic rocks. ***Distribution:*** The upper line of the subalpine belt between 2,350 and 2,500 m; in the Sno gorge, on the Qvena Mt., in the Devdoraki gorge. ***Characteristic species:*** *Sorbus caucasigena, Salix kazbekensis, Rhododendron caucasicum, Vaccinium myrtillus, Anemonastrum fasciculatum, Swertia iberica, Aconitum nasutum, Calamagrostis arundinacea, Dolichorrhiza renifolia, D. caucasica, Cicerbita racemosa, Cephalanthera longifolia.* ***Ecological importance:*** Water-regulatory function, protection of slopes from avalanches, mud-streams, debris-flows, landslides and erosion. ***Threats:*** Tree felling, intensive grazing, construction of roads, climate global change (warming) (Akhalkatsi et al. 2006).

Scrub with dominant *Rhododendron caucasicum*. *Rhododendron caucasicum* – a semi-prostrate evergreen shrub, vegetatively spreading by subterranean creeping stems – is a dominant species in an ecosystem dependent on snow cover. It occurs almost over the whole Caucasus forming dense thickets, in which the cover of this species is usually not less than 90–100 %. *R. caucasicum* forms an endotrophic mycorrhiza, which enables this shrub to successfully colonize poor, acidic soils. ***Habitat:*** Slightly inclined (10–15°) as well as steep (40–70°) north- facing hillsides with deep snow cover in winter. Peat soils of shallow depth (30–40 cm); peat layer located at the depth of 10–15 cm; humus content in the upper soil layer is 23 %. ***Distribution:*** Almost in all the gorges of the region (2,300–2,900 m a.s.l.); individual specimens of *Rhododendron caucasicum* are brought down to 1,800–1,900 m a.s.l. by avalanches (e.g. on the right bank of the river Bidara). ***Characteristic species:*** *Vaccinium myrtillus, V. vitis-idaea, Empetrum caucasicum, Daphne glomerata, Pyrola minor, P. rotundifolia, Anemonastrum fasciculatum, Calamagrostis arundinacea.* Besides the characteristic species noted, the following plants can also be found in the *R. caucasicum* scrub: *Juniperus sabina, Salix* *kazbekensis*. ***Ecological importance:*** Protection of slopes from erosion, avalanches, landslides and mud- and debris-streams; water-regulatory function. ***Threats:*** Intensive grazing, climate global change (warming), uncontrolled collection of leaves and shoots for medicinal purposes (Akhalkatsi et al. 2006).

Low scrub community with dominant *Dryas caucasica*. The areas occupied by these communities are not large. ***Habitat:*** Skeletal and stony, calcareous soils; slaty humid steep (20–50°) slopes with North and North-West aspects. ***Distribution:*** Relatively wide in the Truso gorge and on the Mt. Kuro slopes; between 2,000–2,600 m a.s.l. ***Characteristic species:*** *Deschampsia flexuosa, Daphne glomerata, Vaccinium vitis-idaea, Selaginella helvetica, Primula amoena, Polygonum viviparum, Leontodon danubialis, Parnassia palustris.* ***Ecological importance:*** Protection of humid slopes from erosion. ***Threats:*** Overgrazing, climate warming.

7.3 Biotope Complex: Biotopes of Herbaceous Plants

Subalpine tall herbaceous vegetation (Megaphorbia). The subalpine tall herbaceous vegetation constituted by tall herbs (2.0–2.5 m, rarely 3.0 m tall) is mainly distributed on the western and central parts of the Greater Caucasus, although fragments of this vegetation are scattered all over the Caucasus. Species richness of an individual community is not high. One of the principal characteristics of these plants is their rather rapid development during the growing season. In May and June the vegetation comprises only one layer of low herbs, whilst in the middle of the growing season (the end of July and the beginning of August) it already consists of tall herbs. Tall herbaceous vegetation is typically three-layered; these three layers are constituted not only by one and the same species, but also by one and the same individual plants. Thus, the first layer consists almost completely of generative stems; leaves of the same plants form the second layer, which actually "controls" species content of the community and suppresses the development of other species and individual plants. Strong shading prevents the growth of other autotrophic plants in the third layer except some early spring plants; only stems and elongated leaf stalks are developed in this layer. The main constituents of the tall herbaceous vegetation are forbs; the ecosystem is stenotopic. It is principally distributed in the subalpine belt; however, avalanches often take plants of the community down to lower belts (Dolukhanov 1966a, b; Gagnidze 1974). The main factors in the development of the tall herbaceous vegetation are high air humidity and soil moisture as well as gradual (and not rapid) thawing, slight fluctuations of daily temperatures, frequent fog. ***Habitat:*** Humid north- and west-facing slopes of low inclination (until 10–15°); near subalpine open birch woodlands or in strongly rarefied birch forests; the communities are especially dense on river banks. ***Distribution:*** Between 1,900 and 2,300 m a.s.l. Best developed in the Devdoraki gorge; fragments occur in the Liphu forest, the Sno gorge, near the village Sioni, in the Khde gorge. ***Characteristic species:*** *Heracleum sosnowskyi, Aconitum nasutum, A. orientale, Cephalaria gigantea, Angelica tatianae, Cicerbita macrophylla, Senecio* *rhombifolius, Agasyllis latifolia, Doronicum macrophyllum.* ***Ecological importance:*** Water-balance regulatory function. ***Threats:*** Grazing, tree felling and carry out of firewood, uncontrolled collection of medicinal herbs (especially *Senecio rhombifolius*).

Meadows with dominant *Bromopsis variegata*. These grasslands are widespread. They belong to rare-turf meadows. The dominant species – *Bromopsis variegata* – flowers in the first half of June. The degree of coverage amounts to 95–100 %. ***Habitat:*** Plateaus and south-facing hillsides of low and medium degree of sloping (5–15°) as well as alluvial cones. Moderately dry skeletal soil. ***Distribution:*** Between 1,850 and 2,700 m a.s.l. in the whole Caucasus. ***Characteristic species:*** *Agrostis tenuis, Anthoxanthum alpinum, Festuca ovina, Koeleria luerssenii, Trifolium ambiguum, T. trichocephalum, Ranunculus oreophilus, Alchemilla sericata, Leontodon hispidus, Lotus caucasicus, Platanthera chlorantha.* ***Ecological importance:*** Protection from erosion. ***Practical use:*** Hay-meadows. ***Threats:*** Overgrazing, construction of roads.

Dry meadows with dominant *Agrostis tenuis*. These meadows are widespread in the subalpine belt, but occur in the alpine belt too. Very often *Agrostis tenuis* is a co-dominant of *Bromopsis variegata*. ***Habitat:*** Plateau or south-facing hillsides of medium slope (10–12°). Humid northwest-facing slopes in the alpine belt. Rare-turf meadows. ***Distribution:*** Between 1,800 and 2,600 m a.s.l. in all the gorges and on all the ranges. ***Characteristic species in the subalpine zone:*** *Bromopsis variegata, Festuca ovina, Phleum phleoides, Koeleria luerssenii, Helictotrichon asiaticus, Pedicularis chroorrhincha, Ranunculus oreophilus, R. caucasicus, Trifolium ambiguum, Alchemilla sericata, Gymnadenia conopsea, Coeloglossum viride.* ***Characteristic species in the alpine zone:*** *Poa alpina, Phleum alpinum, Carum caucasicum, Leontodon caucasicus, Taraxacum confusum, Sibbaldia semiglabra.* ***Ecological importance:*** Protection from erosion. ***Practical use:*** Hay-meadows (in the subalpine zone), pastures (in the alpine zone). ***Threats:*** Overgrazing, construction of roads.

Dense-turf xerophilous meadows with dominant *Festuca varia*. These meadows are widespread in the high mountains of the Caucasus. The meadows develop mainly on very steep slopes of any aspect (more often on south-facing slopes). *Festuca varia* is a densely tufted, summer-wintergreen plant (Kimeridze 1965). ***Habitat:*** Very steep (30–50°) stony south-facing slopes. If grazed, they also occur on north-facing slopes. Unlike in the subalpine belt, *Festuca varia* develops in cold and dry environment in the alpine belt, where fluctuations of daily temperatures are sharp. Mountain meadow soil, soddy turf, skeletal. ***Distribution:*** In all the gorges and on all mountain ranges of the region, both in the subalpine and alpine belts. Fragments occur in the subnival belt too. ***Characteristic species in subalpine zone:*** *Helictotrichon asiaticus, H. pubescens, Calamagrostis arundinacea, Oxytropis cyanea, Betonica macrantha, Inula orientalis, Polygonum carneum, Pyrethrum roseum*; ***Characteristic species in alpine zone:*** *Kobresia schoenoides, Carex tristis, Alopecurus dasyanthus, Anthoxanthum alpinum, Alchemilla caucasica, Festuca ruprchechtii, Bromopsis riparia, Polygonum carneum, Primula amoena, Chaerophyllum roseum, Helictotrichon asiaticus, H. pubescens, Podospermum* *alpigenum, Cerastium purpurascens, Betonica macrantha.* ***Ecological importance:*** Protection of slopes from erosion. ***Practical use:*** Grazed well by cattle and satisfactorily by sheep until flowering (while leaves are tender). Hard for grazing after leaves have toughened (Sakhokia 1983). ***Threats:*** None.

Humid broad-leaved meadows with dominant *Trollius ranunculinus*. These meadows are distributed in humid places; they are found in forest openings as well as on slightly sloping hillsides (until 10°) and small depressions of both subalpine and alpine belts. The areas occupied by these meadows are not large. They occur as patches scattered over the other vegetation belts. *Trollius ranunculinus* blooms in May and June. ***Habitat:*** Slopes and small depressions of high as well as medium humidity. ***Distribution:*** Patchy distribution in almost all the gorges of the region; between 1,800 and 2,800 m a.s.l.; these meadows cover relatively large areas on the Kolteshi range, in the Gudauri region. ***Characteristic species:*** *Veratrum lobelianum, Dactylorhiza euxina, D. urvilleana, Poa alpina, Swertia iberica, Deschampsia flexuosa, Pedicularis crassirostris.* ***Threats:*** Extensive grazing, climate global change (warming).

Broad-leaved mesophilous meadows with dominant *Anemonastrum fasciculatum*. These meadows are distributed almost throughout the Caucasus. ***Habitat:*** Moderately humid North and North-West slopes. ***Distribution:*** They occupy quite small areas in the region and develop mainly in the subalpine belt (on the Kolteshi range, in the Oevdoraki gorge) between 2,000 and 2,300 m a.s.l. *Anemonastrum fasciculatum* occurs at forest edges, in *Rhododendron caucasicum* scrub, meadows dominated by *Calamagrostis arundinacea* and *Festuca varia* growing on humid slopes. ***Characteristic species:*** *Trollius ranunculinus, Geranium ibericum, Scabiosa caucasica, Betonica macrantha, Veratrum lobelianum, Polygonum carneum.* ***Threats:*** Extensive grazing, tree felling, climate global change (warming).

Dense-turf meadows with dominant *Nardus stricta*. These meadows are widely spread in the subalpine and alpine belts of the Caucasus. The subalpine meadows made up of *Nardus stricta* are generally secondary, developed as a result of overgrazing (Kimeridze 1965). ***Habitat:*** In level places as well as on hillsides of medium (10–15°) and high degree (20–30°) of slope, on slopes of any aspect (except very steep, south-facing). ***Distribution:*** In the subalpine and alpine belts between 2,000 and 2,800 m a.s.l.; in the subalpine belt the *Nardus stricta* meadows replace the forb-grass meadows, on all the mountain ranges of the region (Sakhokia 1983). ***Characteristic species:*** *Agrostis tenuis, A. planifolia, Luzula pseudosudetica, Anthoxanthum alpinum, Trifolium trichocephalum, T. ambiguum, Leontodon danubialis, Phleum alpinum, Poa alpina, Sibbaldia semiglabra, Hieracium pilosella, Carum caucasicum, Dactylorhiza euxina, D. urvilleana.* ***Ecological importance:*** Protection of slopes from erosion. ***Threats:*** Over-grazing.

Meadows with dominant *Carex tristis*. These meadows are widespread in the alpine belt of the Caucasus. Some fragments of these meadows are also present in the subnival belt. ***Habitat:*** The meadows occupy moderately humid slopes with low (3–5°) and medium (10–15°) degree of inclination (of different aspects) and plane areas. ***Distribution:*** In all the gorges and on all mountain ranges of the region;

between 2,300 and 2,900 m a.s.l. ***Characteristic species:*** *Kobresia capilliformis, Thalictrum alpinum, Poa alpina, Gnaphalium supinum, Nardus stricta, Luzula spicata, L. multiflora, Festuca supina, Anthennaria caucasica, Polygonum viviparum, Alchemilla caucasica.* ***Ecological importance:*** Protection of slopes from erosion. ***Threats:*** Over-grazing.

Meadows with the *Kobresia capilliformis*. These meadows are especially widely distributed in the Central and East Caucasus. The meadows occupy quite small areas in the West Caucasus on limestone. Their invasion of the Caucasus coincides with the xerothermic period after the glacial epoch. ***Habitat:*** Glacial relief, ridges, windy places with thin snow cover; mainly on carbonate soil.

Distribution: Widely distributed between Mt. Kvena Mta and Betlemi Pass, in the Truso gorge and Kolteshi mountain range; between 1,900 and 2,800 m a.s.l. ***Characteristic species:*** *Kobresia persica, Alchemilla elisabethae, Thalictrum alpinum, Polygonum viviparum, Carum caucasicum, Campanula biebersteiniana.* ***Ecological importance:*** Protection of slopes from erosion. ***Threats:*** Extensive grazing.

Mesophilous rare-turf meadows with dominant *Calamaerostis arundinacea*. ***Habitat:*** Level places, slightly inclined (2–5°) as well as steep (20–25°) slopes; mostly within the range of the birch. ***Distribution:*** limited – the Liphu forest, the Sno birch forest, the Devdoraki gorge. ***Characteristic species:*** *Agrostis planifolia, Helictotrichon asiaticus, Deschampsia flexuosa, Geranium ibericum, Anemonastrum fasciculatum.* ***Threats:*** Tree felling, global climate change (warming).

7.4 Biotope Complex: Biotopes of Rocks and Scree

Distinct fluvio-glacial topography, intense present-day denudation process, granitoids bared in consequence of Tergi erosion, black shale and quartzite of Lower Lias age remained uncovered with soil on steep slopes, volcanic lava of the Quaternary with glacial, alluvial and colluvial layers – all these create the environmental conditions necessary for the development of the rock and scree biotopes.

The biotopes of rocks and scree of the Kazbegi region are distinguished by quite high species richness. According to the data reported by Ivanishvili (Ivanishvili 1998), 25 % of the Kazbegi flora is represented by the species of the biotopes mentioned and half of them are stenotopic endemics. It is worth mentioning that there are several mono- and oligotype endemic genera in the flora of the rocks and scree biotopes of the Kazbegi region.

Biotope of moist rocks. ***Habitat:*** Shaded rocks moistened with horizontal precipitation or water flowing down from hillsides. ***Distribution:*** The Devdoraki, Dariali, Kazbegi, Sno, Truso gorges, etc. ***Characteristic species:*** *Parietaria micrantha, P. judaica, Campanula sosnowskyi, C. hypopolia, Diphasium alpinum, Cryptogamma crispa, Polypodium vulgare, Woodsia fragillis, Dryopteris pumila, Primula darialica.* ***Threats:*** Climate warming.

Table 7.5 Species number in the various treeline biotopes of the Central Greater Caucasus (n = 14)

Biotopes	Minimum species number per 25 m^2	Maximum species number	Average species number
Elfin crooked-stemmed forest dominated by *Betula litwinowii*	24	29	27.0 ± 1.6
Shrabbery dominated by *Rhododendron caucasicum*	16	38	25.7 ± 9.3
Low scrub community dominated by *Dryas caucasica*	23	35	30.3 ± 6.4
Tall herb vegetation dominated by *Heracleum sosnowskyi*	6	9	7.5 ± 1.3
Meadow dominated by *Calamagrostis arundinacea*	27	38	31.6 ± 3.9
Meadow dominated by *Festuca varia*	17	43	31.2 ± 7.8
Meadow dominated by *Nardus stricta*	20	47	32.3 ± 7.9
Meadow dominated by *Carex tristis*	25	28	26.5 ± 2.1
Meadow dominated by *Kobresia capilliformis*	22	34	27.7 ± 4.1

Biotope of marly and slaty scree. ***Habitat:*** Dry slaty scree, north-facing slopes. ***Distribution:*** In the subalpine belt, in the whole region. ***Characteristic species:*** *Silene lacera, Erysimum ibericum, E. substrigosum, Linaria vulgaris, L. meyeri, Thalictrum foetidum, Salvia verticillata, Scutellaria leptostegia, Thymus collinus, Bromopsis riparia, B. biebersteinii, Trigonocaryum involucratum.* ***Threats:*** Avalanches and landslides.

Biotope of stones. Rather different plants settle on stones of various origin (glacial, volcanic, etc.); the abundance of the plants growing on stones is often quite high. Many of the species settling on stones are the constituents of the meadow vegetation and communities of rocks and scree. ***Habitat:*** South-facing stones. ***Characteristic species:*** *Sempervivum pumilum, Campanula bellidifolia, Silene ruprechtii, Thymus collinus, Pulsatilla violacea, Festuca ovina, Koeleria caucasica, K. luerssenii, Carex buschiorum, Sedum oppositifolium.* ***Threats:*** Out of danger.

The lowest index of species number is found in the biotope of tall herbaceous vegetation, which is caused by the specific feature of its canopy structure and microclimate (Table 7.5). Other biotopes are characterized by rather high value of species richness.

High-Mountain Vegetation

8

High-mountain vegetation (from the viewpoint of botany) comprises the area situated above the climatic limit of dense forests. They are characterized by specific environmental conditions: high solar radiation, high ratio of ultra-violet radiation, low temperature, intensive pressure of water vapour, etc. (Larcher 1980; Körner 2003). The mean temperature of the growth period is generally 6–8 °C (Körner et al. 2011; Pauli et al. 2011)

One of the most pronounced peculiarities of high-mountains is the diversity of habitats within even a very small area.

All the altitudinal belts characteristic of the Nemoral Zone of temperate climate are represented in the high-mountains of Georgia. The following altitudinal zones can be observed here: subalpine (from 1,800–1,900 up to 2,400–2,500 m), alpine (from 2,400–2,500 up to 2,960–3,000 m), subnival (from 2,900–3,000 up to 3,600–3,750 m) and nival (above 3,700 m a.s.l.).

The majority of Caucasian botanists regarded the high-mountain flora of the region to consist chiefly of the Boreal and Arcto-Alpine elements, which penetrated into this area during the Pleistocene; at the same time they presumed that only some high-mountain species were of autochthonous origin, genetically close to those of the low-mountain zones.

Later, it was stated that the most typical representatives of the Caucasian high-mountain flora are of autochthonous origin (Feodorov 1952; Kharadze 1960). According to these scholars, the high-mountain flora of today, being based on a Tertiary nucleus, developed during the Quaternary, in an autochthonous way.

According to Kharadze (1960), high-mountain landscapes of the Caucasus had been formed before the Pleistocene. The above-mentioned scholar does not share the opinion that the origin of high-mountain flora of the Caucasus is connected with the Late Quaternary. The high-mountain flora of the Caucasus is composed of ancient mesic and xeric elements.

In his analysis of the Caucasian endemism, Grossheim (1936) indicates that the area of the Greater Caucasus, where the "processing" of migrated boreal and Anterior Asian elements took place, is of special significance.

G. Nakhutsrishvili, *The Vegetation of Georgia (South Caucasus)*, Geobotany Studies, DOI 10.1007/978-3-642-29915-5_8,

Kharadze (1960) outlines the following species which have penetrated to the Central Caucasus from Daghestan: *Stipa daghestanica, Thymus daghestanicus, Heliotropium styligerum,* etc. The same scholar suggests that the limestone mountain massifs of the North Caucasus participated in the formation of the pre-Pleistocene xerophytic flora. The example of such mountains is the limestone Rocky Mountain Range. Autochthonous hemixerophytic elements are well represented in the areas of the Central Caucasus from the mid-mountain zone to high-mountains.

Among the distant relations of the Caucasian oreoxerophytic flora its connections with those of the Mediterranean and Anterior Asian should be mentioned. Ancient connections with the Mediterranean flora have been found in the following genera: *Astragalus, Nepeta, Scabiosa, Symphyandra, Edraianthus.* The Late Pliocene is the period of the most ancient migrations from Anterior Asia (= Asia Minor) to the Caucasus. Penetration of these elements into the Caucasus is attributed to the xeric periods of Pleistocene and also of Holocene (Grossheim 1936; Kharadze 1960; Janelidze and Margalitadze 1977).

Autochthonous oreoxerophytic species of the Caucasus reveal ancient connections with those of the Anterior Asian flora. Examples are *Salvia canescens, S. daghestanica, Betonica nivea,* species of *Silene, Astragalus, Scutellaria, Ziziphora,* etc.

Members of the autochthonous high-mountain flora of the Caucasus are regarded as xerophytic elements of the humid flora of the same region (Kharadze 1960).

According to Kharadze (1960), there have been two stages of formation of the Caucasian high-mountain elements. First, in the Miocene-Pliocene, more thermophilous elements had been formed, and after, in the Pliocene-Pleistocene the organisms fully changed in the direction of cryophytization. From the Upper Miocene to the Upper Pliocene, even the most elevated parts of the Greater Caucasus were populated by the representatives of thermophilous flora.

The most conservative palaeoendemics of the Caucasian high-mountains have remained in the mid-mountains and peripheries of the mountain glaciations. In the Alps and Carpathians, as well as in the Caucasus, the extreme mountain ranges served as refuges for the pre-glacial ancient endemic flora (Kharadze 1960). Both progressive and conservative palaeoendemics can be found in the Caucasian high-mountains. Conservative palaeoendemics are exemplified by *Sredinskya grandis, Campanula mirabilis, C. ossetica, Edraianthus owerinianus, Symphyandra lezgina.* Progressive palaeoendemics include such species, as *Charesia akinfievii, Petrocoma hoefftiana* and *Symphyandra pendula.*

Apart from the above-mentioned plants, high-mountains of the Caucasus are populated by the following species (Kharadze 1960):

1. Endemo-vicariants of the Miocene-Pliocene differentiation (*Gypsophila imbricata, Arabis sachokiana, Scabiosa gumbetica, Campanula kolenatiana*);
2. Endemo-vicariants with disjunctive distribution areas (*Ranunculus helenae, Hypericum nummularioides*);
3. Progressive palaeoendemics and highly specialized species of eucryophytic complexes (*Pseudovesicaria digitata, Trigonocaryum involucratum, Pseudobetckea caucasica*);

Table 8.1 Life forms (in a broad sense) of the high-mountain plants

Morphological types
1. *Woody plants*
Straight-trunk-trees: *Acer trautvetteri, Pinus kochiana, Quercus macranthera, Salix caprea*
Crook-stem forest trees: *Betula litwinowii, B. medwediewii, Quercus pontica*
Shrubs: *Rhododendron caucasicum, Rh. ponticum, Rh. ungernii, Rh. smirnowii, Juniperus communis, Rhamnus depressa, Rh. imeretina, Salix kazbekensis*
Dwarf subshrubs: *Daphne glomerata, Vaccinium myrtillus, V. vitis-idaea, Empetrum caucasicum, Dryas caucasica, Salix kazbekensis, Juniperus sabina*
Cushion-shaped shrubs: *Astragalus denudatus, A. microcephalus*
2. *Herbaceous plants*
Plants with upright shoots – *Centaurea cheiranthifolia, Aconitum nasutum, Gadellia lactiflora*
Rosette plants: *Taraxacum stevenii, Chamaesciadium acaule, Plantago caucasica*
Prostrate plants – *Veronica minuta, Minuartia aizoides, Astragalus captiosus, Trigonocaryum involucratum*
Megaphytes – *Heracleum sosnowskyi, Telekia speciosa, Gadellia lactiflora*
Cushion-shaped plants – *Minuartia inamoena, Saxifraga moschata, Draba bryoides*
Creeping plants – *Cuscuta epithymum, Convolvulus lineatus, Vicia grossheimii, V. alpestris*
Types based on relations of the perennating buds to the soil level
Phanerophytes
1. *Macrophanerophytes*: *Acer trautvetteri, Sorbus caucasigena, Betula litwinowii, Quercus macranthera, Salix caprea*
2. *Nanophanerophytes*: *Rhododendron caucasicum, Juniperus hemisphaerica, Rhamnus depressa*
Chamaephytes: *Vaccinium myrtillus, Daphne glomerata, Dryas caucasica*
Hemicryptophytes: *Bromopsis variegata, Festuca varia, Betonica macrantha*
Cryptophytes (e.g. geophytes, hydrophytes): *Fritillaria lutea, Lilium georgicum, Hippuris vulgaris*
Therophytes: All species of *Euphrasia spp, Rhinanthus spp., Gentianella caucasea, Lomatogonium carinthiacum*
Phenorhythmotypes
1. *Evergreen plants*
Continuously foliated plants: *Rhododendron caucasicum, Daphne glomerata, Saxifraga juniperifolia*
Summer-winter-green plants: *Festuca varia, Carex meinshauseniana, Alchemilla sericata*
2. *Vernal-summer-green plants*: *Rhododendron luteum, Betonica macrantha, Aconitum nasutum, Geranium ibericum*
3. *Plants with short vegetative period*: *Fritillaria lutea, Galanthus caucasicus, Gagea glacialis*

4. Endemo-vicariants of the Upper Pliocene differentiation (*Ranunculus lojkae, Cerastium undulatifolium, Cerastium kasbek, Senecio karjaginii*).

High-mountains of the Caucasus and Georgia in particular, display a great variety of life forms, which can be accounted for by their evolution under different climatic conditions, relief dissection, and also by the situation of the Caucasus at the junction of contrasting physico-geographical regions. Table 8.1 represents the classification of life forms.

8.1 Subalpine Vegetation

The subalpine belt involves the areas which are situated above the dense forests up to the lower limit of the prevailing alpine grassland. Thus, the vegetation of the subalpine belt includes not only shrubwoods but also thin crook-stem forests. The subalpine belt manifests remarkably high floristical and phytocoenotical diversity, stimulated by the following relatively optimal climatic conditions:

1. Low daily temperature variations;
2. High air humidity;
3. Rich soils;
4. Not so high (as in alpine and subnival belts) solar radiation.

Due to the climatic conditions, the lower boundary of the subalpine belt varies in altitude between 2,350 and 2,500 m. In eastern and partially in central parts of the Greater Caucasus the upper limit attains 2,450–2,500 m a.s.l.

Subalpine vegetation is well developed in the Caucasus, whereas in the Alps and Carpathians the typical subalpine meadows and tall herbs are comparatively poorly represented; at the same time elfin woodlands become substituted by low grass meadows.

Some botanists regard the subalpine belt as a part of the alpine zone. In the nineteenth and the beginning of the twentieth centuries, subalpine meadows were considered to be of secondary origin.

Later it was stated, that the above-mentioned meadows are examples of climax communities.

In treeless areas of South Georgian Upland, mountain steppes compete with alpine meadows. It is evident that the subalpine belt is not represented in these regions.

The following types of vegetation are well developed in the subalpine belt of the Caucasus and Georgia, in particular:

1. Crook-stem forests;
2. Elfin woodlands;
3. Subalpine tall herbaceous vegetation;
4. Subalpine meadows.

In several regions of Georgia, subalpine the belt is characterized by the occurrence of oreoxerophytic vegetation, as well as meadow steppes and steppes.

8.1.1 Subalpine Tall Herbaceous Vegetation

Unlike many other mountain systems (Alps, Rocky Mts, Himalaya, etc.), tall herbaceous vegetation is well represented in the subalpine belt of the Caucasus. Here it is confined to the lower part of the subalpine zone. Tall herbaceous vegetation is developed under the following favourable environmental conditions:

(a) Optimal air and soil temperature;
(b) Negligible daily fluctuations of temperature;
(c) High air humidity;
(d) Rich soils, etc.

Fig. 8.1 (**a**) *Heracleum wilhelmsii*, Tskhratskaro Pass, Bakuriani region, Lesser Caucasus, (**b**) *Lilium georgicum*, Kazbegi, Central Caucasus (Photo O. Abdaladze)

Tall herbaceous vegetation is composed of tall (3–4 m) herbs, chiefly dicots (Figs. 8.1a, b and 8.2). They are characterized by rosetteless shoots, short top roots and rhizomes. Stratification is not pronounced within the tall herbaceous communities. According to Gagnidze (1974, 1977; Shetekauri and Gagnidze 2000), the total number of species occurring in subalpine tall herbaceous vegetation amounts to 90, while their floro-coenotic complex includes only 70 species (Table 8.2).

A turning point in the composition of the floral-coenotic complex of tall herbaceous vegetation has commenced as early as Middle Pliocene. During subsequent periods, areas with tall herbaceous vegetation underwent reduction and even lost some individual components. According to Gagnidze (1977), regularities of the

Fig. 8.2 *Gadellia lactiflora*, tall herbaceous vegetation, Bakuriani, Lesser Caucasus (Photo A. G. Dolukhanov)

present-day altitudinal distribution of the components of tall herbaceous vegetation of the Caucasus and the final formation of the corresponding floro-coenotical complex are related to the Pleistocene and Holocene.

Dolukhanov et al. (1946) indicate three groups of associations with the following leading species:

1. *Heracleum mantegazzianum, H. grossheimi, H. sosnowskyi,* etc.;
2. *Senecio platyphyllus, Inula grandiflora,* etc.;
3. *Rumex alpinus, Senecio othonnae,* etc.

According to Panjutin (1939), the following associations (plant community) are known from the north-western Colchis (Abkhazeti):

1. Inuletum *(Inula magnifica),*
2. Telekietum *(Telekia speciosa),*
3. Aconitetum *(Aconitum orientale),*
4. Cicerbitetum *(Cicerbita bourgaei, C. cacaliaefolia),*
5. Symphytetum *(Symphytum asperum),*
6. Heracleetum mantegazziani.

The same author attributes Anthriscietum and Campanuletum to intermediate associations (plant community). Weedy tall herbaceous vegetation should be divided into the following seven associations, such as:

1. Delphinietum pyramidati,
2. Senecietum platyphylli,
3. Senecietum chlorocomos,
4. Cirsietum obvallati,
5. Anthriscietum nemorosi,
6. Heracleetum asperi,
7. Pteridietum tauricumi.

We distinguish six formations of tall herbaceous vegetation, as well as several groups and association classes, characteristic of the subalpine belt of Lagodekhi Reserve (Table 8.3).

Table 8.2 Species of tall herbaceous vegetation (Hochstaudenflur) (by Gagnidze 1974)

Aconitum nasutum
A. orientale
Angelica adzharica
A. pachyptera
A. tatianae
Anthriscus nemorosa
Cephalaria gigantea
C. procera
Campanula latifolia
Chaerophyllum maculatum
Cicerbita bourgaei
Cicerbita deltoidea
C. macrophylla
C. olgae
C. petiolata
C. prenanthoides
Cirsium aggregatum
C. albowianum
C. buschianum
C. czerkessicum
C. gagnidzei
C. hypoleucum
C. kuznetsowianum
C. oblongifolium
C. svaneticum
C. sychnosanthum
Delphinium bracteosum
D. buschianum
D. dasycarpum
D. dzavakhischwilii
D. fedorovii
D. flexuosum
D. ironorum
D. mariae
D. megalanthum
D. osseticum
D. prokhanovii
D. pyramidatum
D. speciosum
D. thamarae
Doronicum macrophyllum
Euphorbia macroceras
Gadellia lactiflora
Geranium kemulariae

(continued)

Table 8.2 (continued)

Grossheimia ossica
Heracleum aconitifolium
H. asperum
H. circassicum
H. colchicum
H. cyclocarpum
H. grossheimii
H. mantegazzianum
H. ponticum
H. sosnowskyi
H. trachyloma
H. wilhelmsii
Inula magnifica
Knautia montana
Ligusticum alatum
L. arafoe
L. physospermifolium
Lilium armenum
L. georgicum
L. kesselringianum
L. monadelphum
L. szovitsianum
Milium effusum
M. schmidtianum
Petasites albus
Prenanthes abietina
Pyrethrum macrophyllum
Senecio cladobotrys
S. othonnae
S. platyphylloides
S. pojarkovae
S. propinquus
S. rhombifolius
S. similiflorus
S. subfloccosus
Telekia speciosa
Valeriana alliariifolia
V. colchica
V. tiliifolia
Veratrum lobelianum
Xanthogalum purpurascens

Table 8.3 The phytosociological classification of the subalpine vegetation of the Lagodekhi reservation

Communities
Calamagrostis arundinacea + *Bromopsis variegata*
Calamagrostis arundinacea + *Poa longifolia*
Calamagrostis arundinacea + *Festuca varia*
Calamagrostis arundinacea + *Geranium ibericum*
Calamagrostis arundinacea + *Trollius ranunculinus*
Poa longifolia + *Calamagrostis arundinacea*
Poa longifolia + *Festuca varia*
Poa longifolia + *Agrostis planifolia*
Poa longifolia + *Geranium ibericum*
Poa longifolia + *Betonica macrantha*
Bromopsis variegata + *Calamagrostis arundinacea*
Bromopsis variegata + *Poa longifolia*
Agrostis planifolia + *Trifolium canescens*
Agrostis planifolia + *Alchemilla caucasica*
Inula orientalis + *Geranium ibericum*
Geranium ibericum + *Trollius ranunculinus*
Trollius ranunculinus + *Ranunculus caucasicus*
Festuca varia + *Chamaemelum caucasicum*
Festuca varia + *Betonica macrantha*
Festuca varia + *Nardus stricta*
Festuca varia + *Agrostis planifolia*
Nardus stricta + *Festuca varia*
Senecio platyphylloides + *Milium effusum*
Senecio platyphylloides + *Dactylis glomerata*
Heracleum sosnowskyi + *Telekia speciosa*
Telekia speciosa + *Heracleum sosnowskyi*
Telekia speciosa + *Athyrium filix-femina*
Athyrium filix-femina + *Gadellia lactiflora*
Rhododendron caucasicum + *Geranium ibericum*
Daphne glomerata + *Festuca varia*

8.1.2 Subalpine Meadows

The term "subalpine meadows" was proposed in 1898 by the well-known Caucasian botanist N. Busch. Subalpine meadows comprise all more or less mesic herbaceous communities which occupy the areas above timberline and are characterized by relatively high herbage (Dolukhanov et al. 1942; Nakhutsrishvili 1988). Caucasian subalpine meadows are very diverse floristically, phytocoenotically and display a large spectrum of life forms.

Further, we attempt to characterise the main types of subalpine meadows.

8.1.2.1 Grass Meadows

Bromopsieto (*Bromopsis variegata*); Agrostieta (*Agrostis tenuis, A. planifolia*). These communities occupy both dry and moist localities; they occur on skeletal cobble soils. Main components of the above-mentioned coenoses are *Trifolium ambiguum, Lotus caucasicus, Alchemilla sericata,* etc. The above communities are widespread in the Caucasian high-mountains and reach the altitude of 2,700 m. Tussocks are used as hay-meadows.

Hordeeta (*Hordeum violaceum*). *Hordeum* communities occupy the lower part of the subalpine belt (up to 2,000–2,200 m). Of the various coenoses of this formation special attention should be paid to *Hordeum* meadows with the predominance of such tall herbs as *Heracleum asperum, Anthriscus nemorosa, Seseli transcaucasica* (Table 8.4).

Calamogrostieta (*Calamogrostis arundinacea*). These coenoses occupy wet slopes. They are connected with forests and *Rhododendron* thickets floristically; *Calamagrostis* communities frequently occur on deforested areas. *Calamagrostis arundinacea* appears as a dominant of the herbaceous layer in subalpine oak (*Quercus macranthera*) forests, and a subdominant of variegated fescue *(Festuca varia)* communities.

Of the characteristic associations of West Transcaucasia the following should be mentioned:

1. Calamagrostideto-Betoniceta (*Betonica macrantha*),
2. Calamagrostideto-Geranieta (*Geranium ibericum*),
3. Calamagrostideto-Solidageta (*Solidago virgaurea*).

In western parts of Kavkasioni tussocks dominated by endemic *Festuca djimilensis* occupy both dry and moist localities. Meadows of *Sesleria anatolica* with high presence of *Brachypodium rupestre* and *Carex pontica* are restricted to the limestones of Colchis (Bzipi mountain ridge, Abkhazeti).

Poeta (*Poa longifolia*). These coenoses occupy clearings, forest margins, and wet stony slopes. Table 8.3 includes a list of several communities which are well represented in the Lagodekhi Reserve.

Deschampsieta (*Deschampsia caespitosa*) *Deschampsia* communities occur on pebbles near river banks and swampy places. Common plants of these coenoses include: *Equisetum arvense, Carex canescens, C. hirta, C. irrigua, Parnassia palustris,* etc. (Table 8.5).

Beckmanieta (*Beckmania eruciformis*) communities occur in moist and swampy localities of South Georgian Upland. According to Sh. Nakhutsrishvili (1966), the following bidominant communities populate this area:

1. Phleumeto (*Phleum pratense*) Festuceta (*Festuca pratensis*),
2. Agrostieto (*Agrostis alba*) Dactyleta (*Dactylis glomerata*).

The above-mentioned communities include many plants which are valuable for fodder.

Agrostieta (*Agrostis karsensis*). Communities involved in this group occupy banks of meandering streams in South Georgian Uplands (Nakhutsrishvili 1966).

Table 8.4 *Hordeeta* in Kazbegi

Height (m s. m.)	1,850
Exposition	S
Slope (°)	5
Cover (%)	100
Hordeum violaceum	5
Bromopsis riparia	2
B. variegata	2
Phleum pratense	2
Ph. phleoides	1
Agrostis planifolia	2
A. tenuis	2
Poa pratensis	4
Festuca pratensis	3
Koeleria gracilis	1
Tragopogon reticulatus	2
Rumex acetosa	1
Trifolium ambiguum	3
Myosotis alpestris	2
Rhinanthus minor	+
Lotus caucasicus	+
Pastinaca armena	2
Galium verum	1
Vicia grossheimii	2
Silene vulgaris	1
Trifolium fontanum	2
Ranunculus elegans	2
Seseli transcaucasica	4
Myosotis sylvatica	2
Cuscuta europaea	+
Anthriscus nemorosa	3
Alchemilla retinervis	2
Veronica gentianoides	2
Taraxacum confusum	2
Potentilla recta	+
Draba hispida	+
Leontodon hispidus	2
Pedicularis chroorryncha	2
Carum carvi	2
Euphrasia hirtella	+
Alchemilla sericata	1
Verbascum blattaria	+
Campanula trautvetteri	+
Polygonum alpinum	+
Cerastium arvense	+
Vicia alpestris	+

Table 8.5 *Deschampsieta* in Kazbegi

Height (m s. m.)	1,750
Exposition	W
Slope (°)	3
Cover (%)	80
Deschampsia cespitosa	5
Phragmites australis	4
Festuca arundinacea	2
Festuca pratensis	2
Agrostis planifolia	2
Phleum pratense	2
Carex aequivoca	3
Ranunculus elegans	3
Ranunculus buhsei	2
Rhinanthus minor	2
Cerastium arvense	2
Cerastium caespitosum	2
Linium catharticum	1
Trifolium fontanum	1
Equisetum palustre	4
Trifolium ambiguum	1
Leontodon hispidus	1
Lotus caucasicus	2
Cuscuta europaea	+
Euphrasia hirtella	+
Plantago caucasica	+
Trifolium repens	+
Vicia grossheimii	+
Carum carvi	+
Cruciata glabra	+
Ligularia sibirica	1
Seseli transcaucasica	+
Ranunculus acutilobus	+
Cirsium simplex	2
Geranium ruprechtii	+
Trifolium repens	+

8.1.2.2 Tussocks

***Festuceta variae* subsp.woronowii** occupy huge slopes of subalpine belt, where they form the first stage of succession. Under the influence of intensive grazing, *Calamagrostis arundinacea* communities become replaced by those of variegated fescue, but after grazing is forbidden, demutation of *Festuca* coenoses can be observed. *F. varia* forms large bunches; interspaces are populated by grasses *(Helictotrichon* *asiaticum, H. pubescens, Agrostis planifolia)*, sedges *(Carex meinshauseniana)* and forbs *(Betonica macrantha, Polygonum carneum,* etc.) (Table 8.6).

On wet slopes, *Festuca* communities are richer than those which occur on dry ones.

Jaroshenko (1942) and Geideman (1932) attributed variegated fescue meadows to steppe communities, while Busch (1935) and Magakian (1941) regarded them as typical meadows. Later Jaroshenko suggested, that *Festucetum variae* is a relic steppe, formed during the xerothermal period, and that since then, due to the further increase of climate humidity, the above-mentioned communities have been subject to the process of prairification. This opinion is shared by Grossheim.

Variegated fescue meadows are well-spread throughout the Caucasus. Though in South Georgian Uplands, unlike the Greater Caucasus, these communities occur on grazed areas of northern slopes.

Variegated fescue meadows include such widespread association as Festucetum caricosum *(Carex meinshauseniana).*

***Festuceta valesiacae* steppe meadows** are confined to southern slopes. They are of a relatively simple structure (low herbage, low coverage percent, small number of coenoses). *Koeleria luerssenii, Bromopsis riparia, Agrostis tenuis, Carex buschiorum, Pulsatilla violacea, Thymus collinus* are the main components of these communities.

The driest eroded slopes of the subalpine belt are covered with communities dominated by *Bromopsis riparia, Festuca valesiaca, Koeleria luerssenii,* etc. They grow on thin, cobble and skeletal soils.

According to Tumajaniv (1980), forb-grass meadows observed in park forests of *Q. macranthera* in the eastern part of Kavkasioni, are steppificated. The following species are typical for these meadows: *Brachypodium pinnatum, Bromopsis riparia, Carex buschiorum* and *Onobrychis biebersteinii.*

8.1.2.3 Forb and Grass-Forb Meadows

Meadows of this group are abundant in the areas with moist temperate climate. Florogenetically, these communities are connected with the upper forest belt: they occur in thin park forests. It must be noted, that due to the dominant position of chionophytes, these meadows can rarely be found in the mountain massifs with a complex relief.

Meadows with *Woronowia speciosa* (= *Geum speciosum*) arise particular interest. They are very common in the western part of Kavkasioni (limestone-built mountains of Colchis). Usually, these communities populate the slopes with a sharply defined karst relief. *Woronowia speciosa* coenoses with a considerable admixture of *Carex pontica* occupy the same area.

Woronowia communities present the first stage of succession on karst funnels. Vital activity of *W. speciosa* results in the formation of soil, on which the other meadow elements start to crop up. *Woronowia* meadows are likely to have been more abundant in Colchis before. Due to the influence of grazing, in many areas, *W. speciosa* communities became substituted by those of *Nardus* and alpine carpet-like meadows.

Table 8.6 *Festuceta variae* in Kazbegi

Height (m s. m.)	2,000
Exposition	NW
Slope (°)	45
Cover (%)	100
Carex meinshauseniana	4
Agrostis planifolia	2
Bromopsis variegata	2
Festuca varia	5
Helictotrichon pubescens	2
Calamagrostis arundinacea	2
Poa longifolia	2
Anthoxanthum alpinum	2
Polygonum carneum	3
Vicia alpestris	3
Campanula collina	2
Alchemilla sericata	2
Trifolium ambiguum	1
Primula amoena	1
Betonica macrantha	2
Inula orientalis	2
Pedicularis chroorrhyncha	2
Silene ruprechtii	1
Ranunculus oreophilus	2
Ranunculus caucasicus	1
Pyrethrum roseum	2
Cuscuta europaea	+
Minuartia circassica	2
Minuartia imbricata	2
Scabiosa caucasica	2
Cirsium obvallatum	1
Trifolium trichocephalum	1
Myosotis alpestris	1
Polygala alpicola	1
Leontodon hispidus	2
Thymus nummularius	2
Lotus caucasicus	1
Anemonastrum fasciculatum	2
Alchemilla retinervis	2
Rhinanthus minor	1
Cerastium purpurascens	1
Carum alpinum	+
Centaurea cheiranthifolia	+
Coeloglossum viride	+
Cruciata glabra	+

(continued)

Table 8.6 (continued)

Gentiana septemfida	+
Polygonum alpinum	+
Selaginella helvetica	+
Viola odorata	+

Table 8.7 Subalpine *Galanthus* Meadow (grazed) near Jvari pass on the Georgian military highway over the central greater Caucasus (relevé G-10) (Box et al. 2000) (Location: Jvari Pass (2,394 m), Georgian Military Highway, south side (27 May 1999))

			2,300 m, Slope: 5° to SEE	
H	0.3 m	75 %	10 × 10 m KF, GN, *EB*, RJL	
H	3.4	*Galanthus platyphyllus*	3.3	*Alchemilla retinervis*
	2.2	*Rumex alpinus*	2.2	*Festuca supina*
	2.2	*Phleum alpinum*	2.2	*Rumex acetosa*
	1.2	*Sedum oppositifolium*	1.2	*Gagea supranivalis*
	1.2	*Veratrum lobelianum*	1.2	*Cirsium obavallatum*
	1.1	*Fritillaria latifolia*	+.2	*Ligusticum alatum*
	+.2	*Lamium album*	+.2	*Poa alpina*
	+.2	*Geranium ruprechtii*		

Meadows of *Trollius patulus* are very common in forest margins, thin park forests and elongated cavities.

Ranunculus caucasicus frequently occurs in these communities as a subdominant. Pure *Ranunculus* meadows have been formed under the influence of overgrazing (Grossheim 1948).

Geranieta (*Geranium gymnocaulon*) communities are confined chiefly to the alpine belt of the western part of the Greater Caucasus, though they also come down to the subalpine belt.

The distribution area of ***Geranium platypetalum*** meadows comprises Kavkasioni and the Lesser Caucasus. They are unavailable in South Colchis. These communities occur on stony substratum of moraines. With the development of soil cover, *G. platypetalum* communities are being replaced by other communities.

Forest margins, mountain slopes of medium steepness and flat areas are inhabited by *Geranium ibericum*, although as subdominant this species more frequently occurs in Calamagrostideta, Inuleta, Rhododendreta, etc.

Table 8.7 shows a sample of a (grazed) subalpine *Galanthus-Alchemilla* meadow near the Jvari Pass (2,300 m) on the Military Highway over the Greater Caucasus. The species total is again somewhat small, but the species are interesting local species from familiar Eurasian genera. At Kazbegi, at about 1,800 m on the north slope of the Greater Caucasus (Box et al. 2000).

Inuleta (*Inula orientalis*) coenosis are represented by pure thickets of *Inula*. It is very common in the high-mountains of the Caucasus. The following communities are characterized by predominance of this species:

Fig. 8.3 *Anemonetum fasciculatae*, Tskhratskaro Pass, Bakuriani region, Lesser Caucasus (Photo O. Abdaladze)

Inuletum geraniosum *(Geranium ibericum)*,

Inuletum betonicetosum *(Betonica macrantha)*.

In the western part of the Greater Caucasus, communities of *Inula grandiflora* usually occur on negative landforms where they replace tall herbs (Kolakovsky 1961). *Astrantia pontica* coenoses occupy screes and cobble substratum in limestone mountain ranges of Colchis (Abkhazeti).

Of the forb meadows, which are very rich in species, *Scabiosa caucasica* should be noted. These communities include such plants as *Helictotrichon pubescens, Inula orientalis, Agrostis planifolia, Betonica macrantha,* etc.

These meadows are very common in the Lesser Caucasus.

Betoniceta (*Betonica macrantha*) is a typical variant of subalpine meadows. It is well represented in the central and eastern parts of Kavkasioni and the Lesser Caucasus; the above-mentioned communities are rarely found in the western part of the Greater Caucasus.

Anemoneta (*Anemonastrum fasciculatum*) frequently populate slopes with optimal moisture and flat sites (Fig. 8.3). Slopes exposed to sunlight and cobble sites are occupied by communities of hemixerophytic *Hedysarum caucasicum.*

Subalpine meadows of *Veratrum lobelianum* are abundant in western and partly Central Kavkasioni, as well as in the Lesser Caucasus. *Veratrum lobelianum* is a poisonous pasture weed, usually rejected by animals; it is developed on the background of forb (*Trollius ranunculinus* (Fig. 8.4), *Ranunculus caucasicus,* etc.) meadows.

Pulsatilleta (*Pulsatilla violacea*). These communities, typical for negligible areas, are confined to the slopes of northern exposure. *P. violacea* is frequently associated with hemixerophitic grasses (*Festuca ovina, Koeleria luerssenii*) and *Carex buschiorum* (Table 8.8).

Fig. 8.4 *Trollius ranunculinus*, Kazbegi, Central Caucasus (Photo G. Nakhutsrishvili)

Pulsatilleta (*Pulsatilla aurea*). These communities are very common in north-western Colchis. The distribution area of *P. albana* is more expansive.

Geranieto (*Geranium gymnocaulon*) – Woronowieta (*Woronowia speciosa*), Geranieto-Inuleta (*Inula magnifica*) illustrate subalpine mixed forb meadows, which often consist of 2–3 species and are found in Colchis. Pure forb communities are formed by *Euphorbia oblongifolia, E. scripta,* etc. Grass-forb meadows display high coenotical and floristic diversity. According to Kolakovsky (1961), the following species are very abundant in the above-mentioned meadows of Colchis: *Aquilegia olympica, Psephellus abchasicus, Kemulariella caucasica, Senecio aurantiacus,* etc.

Grossheimia polyphylla (*Asteraceae*) is a typical component of grass-forb meadows which are confined to the Lesser Caucasus and partly to western Kavkasioni.

The following plants frequently occur in subalpine grass-forb of the Caucasus: *Centaurea fischeri, C. cheiranthifolia, Vicia alpestris, Lotus caucasicus, Veronica gentianoides, Trifolium ambiguum, Leontodon hispidus, Ranunculus oreophilus,* etc.

Communities dominated by *Polygonum carneum* are worthy of notice; their area of distribution involves Kavkasioni and southern uplands. These meadows are of particular importance as natural grasslands. Under the influence of overgrazing, forb and grass-forb meadows become replaced by more resistant communities of compact-tussock grasses.

Astragaleta captiosus is characteristic of subalpine belt of the Central Caucasus and especially of Kazbegi region; these communities, formed by endemic *Astragalus captiosus*, occupy stony and pebble substratum.

8.1.3 Tragacanthic Vegetation

On the northern slopes of Kavkasioni, tragacanthic vegetation is sporadically distributed in dry inter-mountain vallies and hollows (Ivanishvili 1973).

Table 8.8 *Pulsatilleta* in Kazbegi

Height (m s. m.)	2,050
Exposition	S
Slope (°)	35
Cover (%)	90
Festuca rupicola	5
Carex buschiorum	4
Bromopsis variegata	2
Agrostis tenuis	2
Koeleria cristata	2
Koeleria caucasica	2
Bupleurum polyphyllum	2
Plantago caucasica	3
Pulsatilla violacea	4
Anthyllis variegata	2
Potentilla crantzii	2
Trifolium ambiguum	1
Ranunculus oreophilus	2
Selaginella selaginoides	1
Leontodon hispidus	2
Silene ruprechtii	1
Cerastium arvense	2
Veronica gentianoides	2
Thymus nummularius	2
Alchemilla sericata	2
Ranunculus acutilobus	1
Campanula collina	2
Euphrasia hirtella	1
Lotus caucasicus	1
Minuartia oreina	1
Ranunculus buhsei	+
Astragalus captiosus	+
Minuartia circassica	+
Pedicularis chroorrhyncha	+
Rhinanthus minor	+
Thesium procumbens	+
Campanula bellidifolia	+
Gentiana aquatica	+

Caucasian tragacanthic vegetation is a version of Asia Anterior–South-West Asia mountain-xerophytic vegetation; these communities are very rare in the Mediterranean mountains.

Astragaleta denudati (*Astragalus denudatus* (Fig. 8.5)) in Khevi (Central Caucasus) is restricted to the lower part of subalpine belt. These communities alternate with steppe-like meadows and saxicolous groupings; they occupy areas

Fig. 8.5 *Astragalus denudatus*, Kazbegi, Central Caucasus (Photo O. Abdaladze)

once populated by pine forests. In Daghestan (Eastern Caucasus) the above-mentioned coenoses can also be met in alpine belt; some individuals of *Astragalus aureus* penetrate even into the subnival zone (up to 3,150 m; Prima 1974).

Tragacanthic communities are dominated by plants, characterized by spiny cushion-like form. These communities also include the following plants:

- Dwarf subshrubs (species of *Artemisia, Thymus, Scutellaria*);
- Mountain steppe grasses (*Elytrigia gracillimum, Bromopsis riparia, Stipa tirsa,* etc.);
- *Carex buschiorum*;
- Hemixerophytic shrubs (*Juniperus hemisphaerica, J. sabina, Rhamnus tortuosa*).

The flora, formed on rocks has been enriched with lithophytes and chasmophytes, which involve many local endemics (Ivanishvili 1973).

The following is a concise list of species characteristic of one of tragacanthic communities:

Astragalus denudatus	*Stipa caucasica*
Berberis vulgaris	*Stipa tirsa*
Juniperus hemisphaerica	*Alopecurus vaginatus*
Spiraea hypericifolia	*Carex buschiorum*
Ephedra procera	*Oxytropis cyanea*
Artemisia chamaemelifolia	*Allium albidum*
Artemisia marschalliana	*Allium ruprechtii*
Artemisia splendens	*Asperula albovii*
Scutellaria leptostegia	*Astragalus kazbeki*
Agropyron gracillium	*Campanula hohenackeri*
Bromopsis riparia	*Dianthus cretaceus*

(continued)

Festuca valesiaca	*Onosma caucasica*
Koeleria cristata	*Myosotis arvensis*
Melica transsilvanica	

8.1.4 Microclimate and Energy Balance

A characteristic feature of high mountain ecosystems is their considerable variability within a relatively small area. As a result of particularly harsh climatic conditions (especially directed ones, like solar radiation and wind) and the relief of the high-mountains, a mosaic of various microhabitats emerges.

The microclimate and energetic conditions of phytocoenoses have been studied (by the method of Cernusca 1976) on the following objects, selected for this purpose:

- *Hordeeta*: (1,850 m a.s.l.) The meadow represents the *Hordeum violaceum-Poa pratensis-Anthriscus nemorosa* community with a very rich composition, including up to 70 species (Table 8.4).
- *Festuceta*: (2,000 m a.s.l.) The pasture meadow occupies extremely slanting and rocky slopes of a south-western exposure. The *Festuca varia-Carex meinshauseniana* community is widespread in the subalpine and alpine belts of the Central Greater Caucasus (Table 8.6).
- *Deschampsieta*: (1,750 m a.s.l.) These hay meadows are found on well-moistened habitats. Dominating species are *Deschampsia cespitosa* and *Equisetum palustre*, with regular occurrence of *Phragmites australis* (Table 8.5).
- *Pulsatilleta*: (2,050 m a.s.l.) Is part of an intensively grazed vast pasture. This is *Pulsatilla violacea-Festuca rupicola-Carex buschiorum* community (Table 8.8).
- *Dryeta*: (2,050 m a.s.l.) The community is found on an extremely slanted north-western slope. As a result of intensive grazing, the substrate is characterized as "step" relief. The dominants are *Dryas caucasica*, *Deschampsia flexuosa* and *Daphne glomerata* (Table 8.13).
- *Heracleeta*: (2,200 m a.s.l.) This tall herbaceous phytocoenosis is found on gently slanting northern slopes, in a highly humid substrate. Dominating species, *Heracleum sosnowskyi*, grows 2.4 m in height. The remaining seven species are rather scarce.
- *Kobresieta*: (2,175 m a.s.l.) Is found on gently slanting north-western slopes. The dominants are *Kobresia capilliformis*, *C. persica* and *Carex buschiorum* (Table 8.11).
- *Rhododendreta*: (2,200 m a.s.l.) Is found on steep north-western slopes. It is *Rhododendron caucasicum* and *Vaccinium myrtillus* community. The former is more developed (Table 8.12).

Analysis of the canopy structure and spatial orientation of leaves; microclimatic characteristics, profiles of soil, canopy and air temperature and humidity, falling

Table 8.9 Distribution of photosynthetically active radiation (RhAR) in different plant communities (Tappeiner et al. 1989)

Community	Recorded radiation %	Albedo %	Photosynthesizing parts of plants (leaves, green, shoods, etc.) %	Non-photosynthesizing parts of plants (trigs, dead matter, etc.) %	Soil %
Hordeeta (*H. violaceum*)	100	5	84	10	1
Festuceta (*F. varia*)	100	4	30	65	1
Deschampsieta (*D. cespitosa*)	100	5	84	5	6
Festuceta (*F. ovina*)	100	7	53	14	26
Dryeta (*D. caucasica*)	100	7	65	25	3
Heracleeta (*H. sosnowskyi*)	100	5	88	7	0
Rhododendreta (*Rh. caucasicum*)	100	3	92	5	0

and reflected radiation, net radiation, soil and convective heat flows and evapotranspiration have been made on the previously measured sites, which are described above.

As a result of this investigation of the structure of phytomass, the incident angle and the spatial orientation of leaves, three types of canopy have been relieved (Tappeiner et al. 1989).

1. Canopy with uniform leaf orientation (on different slopes), the phytomass concentrated in the lower parts of canopy.
2. Canopy with uniform distribution of phytomass in height and with basically erectophile leaf orientation in all canopy layers.
3. Canopy with a "storey" distribution of the phytomass and mainly planophile leaf orientation. Individual plant species are dominant.

Dryeta, Pulsatilleta and Kobresieta can be attributed to the first type of canopy. The basic layer of radiation exchange in these phytocoenoses is close to the ground. The second type involves Festuceta and Deschampsieta with low absorption of radiation, observable in the upper layers of the phytocoenosis and almost linear absorption of radiation in the lower part of the canopy. Hordeeta, Heracleeta and Rhododendreta communities, which belong to the third type, reveal absolutely different strategy of adaptation to radiation. The bulk of the biomass and the assimilative ingredients are accumulated in the upper layers of the canopy, thus providing maximum utilisation of the PhAR by their assimilating organs.

As shown in Table 8.9, intensively grazed phytocoenoses are the least utilizers of falling PhAR (Hordeeta pasture; Pulsatilleta; Kobresieta).

Measurements of vertical temperature profiles prove that irregularities between the temperature of the canopy and that of surrounding air are being smoothed.

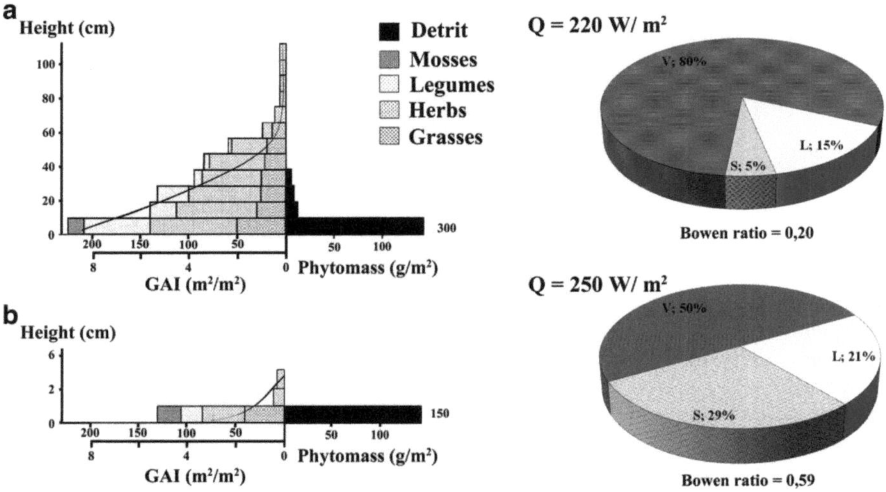

Fig. 8.6 Canopy structure and energy budget in Hordeetum community

Regarding the energy balance of these phytocoenoses, a similar (post-grazing) tendency of utilisation of the absorbed solar energy into heat exchange and transpiration is evident. On all pastures, the ratio of evapotranspiration (20–40 %) in the net radiation is reduced, while the ratio of convective and soil heat flows considerably grows demonstrates the canopy structure (Fig. 8.6), PhAR absorption and the energetic pattern of Hordeeta – both in preserve and on pasture. All these obvious modifications (often irreversible) have been caused by strong human impact.

Every high-mountain phytocoenoses in its natural development, acquires a special strategy of adaptation to environmental conditions, expressed in a definite spatial canopy structure (leaf orientation, distribution of LAI, etc.), which provokes a specific course of radiation energetic processes within the canopy. Heavy exploitation of the phytocoenoses (extensive haymaking, grazing) may stimulate irreversible processes, destroying habitual organisation of the canopy, leading to alterations in the direction of successive processes and moreover, arise dangerous erosion processes in the complex high-mountain relief. Therefore, exploitation of community should be subject to time limitation so as to prevent any such consequences and also to preserve the unique originality of high-mountain ecosystems.

8.1.5 Water Relations

An integral part of investigations into plant ecology is the study of the most important indicator of the vital activity of plants water relations. Water relations in the high-mountain plants of the Central Caucasus have been subject to intensive and long-term observations since the beginning of 1960s. Results of the analysis have been published in Georgia and abroad. However, the purpose of our further investigations is not only to expand the spectrum of the areas studied, but also to reconsider certain problems of water relations in high-mountain vegetation in view of availability of highly sensitive and compact field measuring devices and owing to new conceptions in plant ecology.

Our investigations confined the thesis that the indices of water relations in plants, i.e. water contents in leaves, xylem water potential and leaf transpiration, studied in order to determine their characteristic features, present three interrelated and complementary physiological parameters, giving an accurate picture of the water relations in plants. The determination of the diffusion resistance of leaves allowed us to judge of the function of stomatal apparatus in this vital process (Körner and Cernusca 1988). In order to obtain structural characteristics of water relation, we had to study various anatomical and morphological peculiarities of leaves; for ecological purposes – the phytoclimate and other environmental factors, controlling water relations. The soil moisture has also been carefully examined.

Particular attention was paid to the analysis of regulation mechanisms of the water relations in high-mountain meadow plants. By recent observations even in the subalpine belt of the Central Caucasus, where hydrothermal conditions are most favourable, water evaporation in plants is frequently magnified according to the considerable rise of air temperature and PhAR in the afternoon. Consequently, water contents and xylem water potential noticeably diminish. These conditions stimulate leaf diffusion resistance, hence normal water supply is quickly restored (Fig. 8.7). It should be noted, that physiological regulation by transpiration is expressed rather differently in species of various ecological groups and life forms.

Stomatal regulation of water exchange in plants of the subnival belt of the central Caucasus is much more pronounced. Due to extreme ecological conditions, plants of this belt sharply diminish water release in the afternoon (88–97 %) reducing transpiration to insignificant values as a result of intensification of leaf diffusion resistance.

However, the most important part of the stomata apparatus of leaves in the regulation of water relations of subalpine meadow plants becomes more evident by the end of summer, during the late-summer phase of seasonal development of plants in the course of long-term (10–15 days) drought periods, when the soil is almost devoid of moisture.

During this period the value of leaf moisture and xylem water potential considerably decrease. This situation provides sharp diurnal fluctuations of leaf diffusion resistance, followed by sudden (55–70 %) expansion of xylem water potential. Consequently, plants restore their normal level of tissue water contents (Fig. 8.8).

Considering the ecological fragility of high-mountain biomes, various aspects of the impact of human factor arise particular interest. On the basis of studies on the effect of grass-mowing and grazing on water exchange in high-mountain plants, the most favourable exploitation regime applicable to subalpine meadows has been determined, i.e. a systematic, regular (2–3 times per year) grass-mowing. Grazing or complete reservation, which cause gradual reduction of the green phytomass, seem inexpedient.

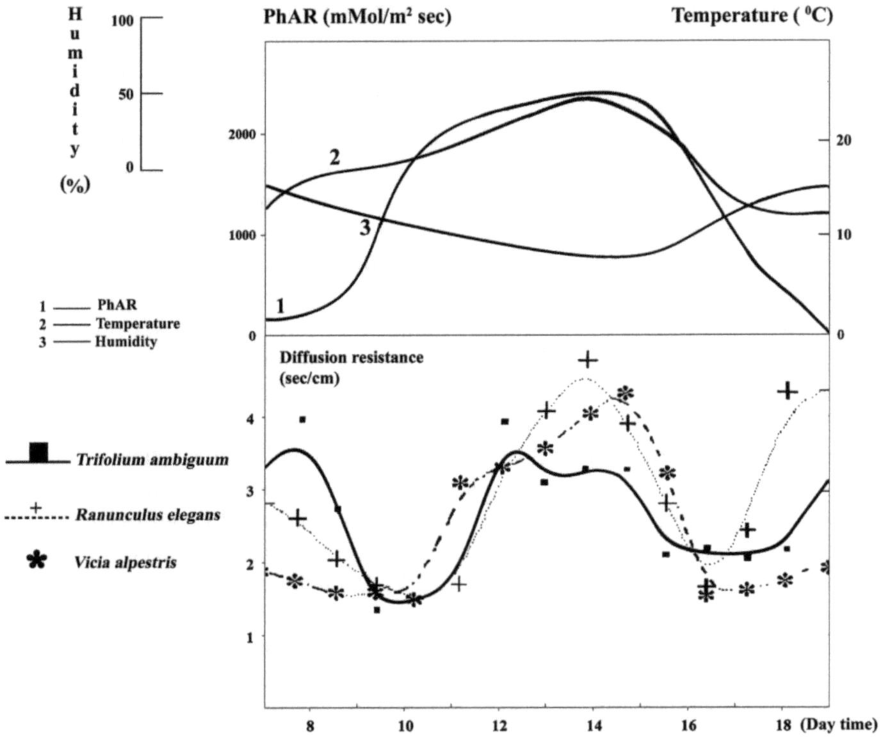

Fig. 8.7 Diurnal course of leaf diffusion resistance of subalpine plants (Sanadiradze, 1986)

8.1.6 Pigment Content

Content of chlorophyll and carotinoid pigments in leaves of more than 100 species of Central Caucasus plants has been subject to analysis in various habitat conditions at the altitudes of 1,800–3,000 m. The pigment content and distribution in different layers of phytocoenoses has been carefully studied. Chlorophyll content in leaves was determined spectrophotometrically (Khetsuriani, Chiboshvili 1998).

By their chlorophyll and carotinoid content, the examined species clearly differ from each other as to their habitat and phytocoenoses, so within a single phytocoenosis. Results of our investigations are illustrated here by the variegated fescue community, found in moderately humid habitats of the north-western slope exposure (1,950 m a.s.l.).

We selected 20 most typical species out of 60 plant species from this phytocoenosis, for close observation during the flowering phase.

Low chlorophyll content (2.66 mg g of dry mass) is typical for the dominant edificator *Festuca varia*, and the maximum (11.8 mg) – for *Vicia alpestris*. The low content of chlorophyll in dominant species can be accounted for by almost erectophyile orientation of leaves, unshaded by other plants.

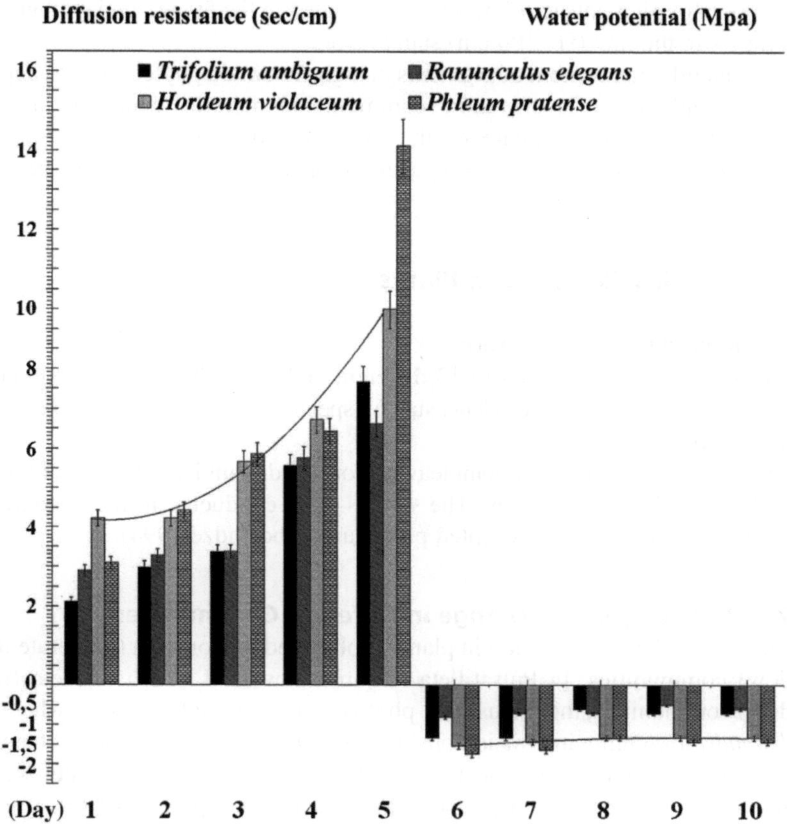

Fig. 8.8 Oscillation of plant water potential

Spring-flowering plants (*Anemonastrum fasciculatum, Primula amoena, Fritillaria lutea*) and other plants of variegated fescue community start their pigmentation synthesis when still under snow. They differ very little by chlorophyll content and the quantity of carotinoids. Though vernal plants develop in heavy ecological conditions, they accumulate as much pigment during their flowering phase as codominant *Carex meinshauseniana*, and twice as much as *Festuca varia.*

Comparison of the quantity of chlorophyll (mg per sq.m) and green mass (g. sq.m) in each species reveals a direct relation between chlorophyll content and biomass production in the majority of these species.

Variability of communities is also determined by their pigment content. Hence the difference between hay and pasture meadows.

Allocation of pigments depends on the height of phytocoenosis, e.g. pigment measurement in every 10 cm layer of the stand (herb-barley meadow) shows that max. amount of pigments (5.75 mg) is accumulated in the 50–60 cm layer and min. (1.00 mg) – in the first (90–100 cm) stand layer.

Consequently, the content of pigments in high-mountain phytocoenoses depends on the complexity of vegetation community structure, which in its turn predetermines the microclimate of the given phytocoenosis. Further, pigment content displays specific variety as a genetic property, in terms of plant species.

8.1.7 CO_2-Gas Exchange in Plants

8.1.7.1 Materials and Methods

Investigations were carried out in 13 different and most characteristic habitats of the Central Greater and Lesser Caucasus. 61 species of flowering plants have been studied for this purpose.

CO_2-gas exchange of intact plant leaves proceeded in under field conditions with infrared gas analyser techniques. The studies were conducted in the open system according to the universally accepted procedure (Abdaladze 1994).

8.1.7.2 Plant CO_2-Gas Exchange in Different Communities

Most intensive CO_2-assimilation in plants is observed in Hordeeta (moderate moist meadow) communities. In Pulsatilleta (dry meadow) and Deschampsieta (moist meadow) communities, the intensity of photosynthesis is slightly lower, whereas in *Rhododendron* communities it is considerably suppressed. CO_2-gas exchange in plants in the communities Festuceta (except *Festuca varia*) on dry south-western mountain side is strongly depressed. Particularly, net-photosynthesis midday depression frequency, duration and capacity are substantially increased and net-primary productivity-decreased.

CO_2-gas exchange in *Festuca varia* (tussock grass with rigid and rolly peinomorphic leaves) is very steady: net-photosynthesis midday depression frequency is extremely low (5–8 %), as well as the coefficient of variation of CO_2-gas exchange (22–25 %). It should be noted, that all parameters in this species display an amazingly constant character on moderately moist slope, on relatively dry slope and at the upper boundary of distribution area of these species (3,000 m a.s.l.), where *Festuca varia* is not a dominant species and is represented by isolated specimen. The CO_2-gas exchange character in *Festuca varia* can be considered as the ecophysiological basis of increased biological activity of this strong dominant, that promotes (particularly in the anthropogenic conditions) its intrusion into different coenoses – "capture" and "holding" of various territories.

Succulent species *Sedum oppositifolium* and *Sempervivum pumilum* are highly adapted to extreme temperature relations of their microhabitats. These succulents assimilate CO_2 by the CAM pathway, which is the most economical way of adaptation for their CO_2-gas exchange.

C3-type of carbon assimilation is typical for *Saxifraga juniperifolia* and *Sedum gracile*.

The obtained data shows that in winter, the evergreen shrub *Rhododendron caucasicum* blocks maximally the interaction with environment and entirely restricts metabolic processes. Net-photosynthesis is not detectable. Dark respiration rate is extremely low (0.2–0.5 mmol CO_2/m^2.s). In winter this species undergoes "deep resting" period.

Contrary to evergreen shrubs, summer-wintergreen herbs (*Plantago lanceolata, Phleum pratense, Trifolium alpestre, Carex sylvatica*) in winter are capable of CO_2 uptake. During thaw the diurnal balance of CO_2-gas exchange, is usually positive. Maximum values of net-photosynthesis on the plots free from snow reach 5.4–9.5 mmol CO_2/m^2.s (37–50 % of vernal maximum). Under the snow cover CO_2 uptake is possible only for a short period of time (1–3 h a day).

8.1.7.3 Maximum Net-Photosynthesis

High net-photosynthesis maximum values show: the legumes – *Trifolium ambiguum, Trifolium fontanum, Vicia grossheimii,* etc. (17–20 mmol CO_2/m^2.s), *Ranunculus oreophilus, Pulsatilla violacea, Leontodon hispidus, Alchemilla sericata, Geranium gymnocaulon, Sibbaldia semiglabra* (18–22 mmol CO_2/m^2.s). The grasses (*Agrostis planifolia, Hordeum violaceum, Bromopsis riparia, Nardus stricta, Festuca ovina, F. varia,* etc.) show moderate values (10–15 mmol CO_2/m^2.s). Low maximums are characterized by C3 succulents, *Rhododendron caucasicum, Parnassia palustris* and the sedges (4.5–10 mmol CO_2/m^2.s).

8.1.7.4 Dark Respiration

High intensity of dark respiration at standard temperature (10 °C) is typical for meadow forbes and legumes (1.7–2.3 mmol CO_2/m^2.s). Sedges and C3-succulents reveal low values (0.9–1.3 mmol CO_2/m^2.s). Grasses show moderate values.

8.1.7.5 Influence of PhAR and Temperature on Plant CO_2-Gas Exchange

Typical shade-tolerant species are: *Saxifraga juniperifolia* and *Sedum gracile* (PhAR saturation intensity =450 mmol photons/m^2.s), *Betonica macrantha and Rhododendron caucasicum* (560–580 mmol photons/m^2.s). The most light-requiring species are: *Pulsatilla violacea, Trifolium ambiguum, Geranium gymnocaulon, Sibbaldia semiglabra* and some others (1,000–1,500 mmol photons/m^2.s; 42–48 % from full PhAR).

The majority of the investigated species are capable to assimilate CO_2 at negative air temperature (−4 to 0 °C). Maximum temperature limit for net-photosynthesis in subalpine and alpine meadow plants is rather high (37–44 °C). Wide temperature optimums for net-photosynthesis in *Festuca varia* (12–13 °C), *Saxifraga juniperifolia* (11–30 °C), *Sibbaldia semiglabra* (16–34 °C) and *Geranium gymnocaulon* (15–34 °C) attract special attention.

The ratio Q_{10} for dark respiration of investigated species equals approximately 2.0. At very high temperature (40 °C and more Q_{10} falls to 1.3–1.4).

8.1.7.6 Stress Factor in High Mountain and Plants CO_2-Gas Exchange

High mountains are stress dominated areas. The strongest negative effect, produced by the stress factor is a destroyed balance between photosynthesis and respiration. Reduction of the value photosynthesis/respiration in investigated species is caused by the increase of altitude above sea level. In heavy ecological conditions plants spend more energy on survival and adaptation.

Obviously, some herbs and shrublets in Rhododendreta become subject to stress caused by reduced PhAR. In cloudy weather relative illumination of leaves lowers to 0.5–1 % so that plants fail to assimilate CO_2.

By the end of spring, frosts reduce net-primary productivity in plants in Agrostieta on 80–85 %. In Festucetum (dry south-western slope) in all investigated species, except *Festuca varia*, net-primary productivity decreases for 11–35 %, because of high air temperature. Draught is a stress factor even for the plants growing under temperature conditions. The stronger is the draught, the greater is the influence of water stress on CO_2-gas exchange in *Hordeum violaceum* in Hordeeta.

As a result of thaw on the south slopes the photoinhibition, the freezing of plants at night and the "winter draught" phenomenon are observed. Therefore, strong and long-term thaws, provoking the activation of net-photosynthesis in herbs should be considered as unfavourable event.

No universal mechanism of plant CO_2-gas exchange adaptation to high mountain ecotope was revealed. CO_2-gas exchange shows the best adaptation to the main limiting ecological factor for a given biotope (Abdaladze 1994).

8.1.7.7 Phytocoenoses Effect on Plant CO_2-Gas Exchange

The survivability of canopy can be revealed experimentally: the protective effect of canopy on CO_2-gas exchange in *Trifolium ambiguum* has been investigated in Hordeetum. Separation, e.g. removal of vegetation around the experimental plants within the radius of 1 m, greatly effects the dependence of net-photosynthesis on PhAR and leaf temperature, suddenly narrowing ecological optimums of these factors. On sunny days for separated plants the highest values with negative index of correlation rate (net-photosynthesis-PhAR $= -0.93 \pm 0.14$; net-photosynthesis-leaf temperature $= -0.90 \pm 0.17$) are observed. It indicates the linear dependence on the net-photosynthesis depression from the illumination and leaf temperature. The correlation interactions of these dependencies are equal and this confirms the linear character of connection. In natural canopy the correlation rates have positive index. Moreover, it smoothed out the temperature fluctuations, which is expressed by net-photosynthesis ratio of variation reduction. Private correlation rate between leaf temperature and net-photosynthesis is approximately zero. It indicates, that canopy not only smoothes up the unfavourable factors but creates actively the optimal conditions for CO_2 uptake. Thus, temperature optimum in control plants equals to 17–26 °C, while an experimented one – 17–23 °C. Separation induces twice and a half times decrease of the net-photosynthesis maximums frequency. It also demonstrates the ability of vegetation cover to support favourable conditions temperature, air humidity, PhAR, etc.

This effect is directed against stress factors of the environment for support of optimal CO_2-gas exchange balance and is probably the result of a cooperative functioning of plants formed in the process of a long-standing joint evolution.

8.1.7.8 Anthropogenic Influence of Plant CO_2-Gas Exchange

Our investigations have been carried out in Hordeeta in the pasture, the hay and preserved plots. On intensively grazed meadow CO_2-gas exchange of plants is destabilised. Here, net-photosynthesis strong midday depressions are observed rather clearly (depression rate = 54–59 %). On preserved and hay meadows net-photosynthesis reduction at depressions does not exceed 38–45 %, whereas on pastures it is 78–81 %. On pastures, net-primary productivity and diurnal balance of CO_2-gas exchange are substantially depressed.

On hay meadows (once a year) optimal CO_2-gas exchange balance was observed (Abdaladze 1994).

8.1.8 Life Activity of High-Mountain Plants in Winter

Life activity of high-mountain plants in winter has been studied in the upper-forest and subalpine belts of the Lesser Caucasus. Our objective was to examine the strategies of hibernation of summer-winter-green meadow, herbaceous and ever-green shrub plants.

On the basis of the obtained data, winter evergreen shrubs usually hinder any interactions with the environment and entirely restrict metabolic processes. Therefore, leaf and air temperatures virtually remain immutable, water content in leaves slightly changes (1–2 %) compared to autumn, stomata are closed, transpiration and net-photosynthesis are imperceptible, respiration rate is extremely low. In profound dormancy, resistance of shrubs against various stress factors is much higher. Safe hibernation depends on snow protection, which prevents plants from desiccation and dehydration.

Contrary to evergreen shrubs, summer-winter-green herbaceous species are capable of vital activity at ever favourable opportunity in winter, particularly during thaw.

In free of snow areas, leaf temperature of the hibernating species amounts to high values on clear days, exceeding air temperature by 5.0–9.3 °C. Diurnal dynamics of water content, xylem water potential, transpiration intensities and net-photosynthesis are displayed rather clearly. During thaw the diurnal balance of CO_2-gas exchange is usually positive, whereas maximum intensity of the net-photosynthesis attains 37–50 % of vernal maxima. CO_2 assimilation under snow cover becomes possible only for a short period.

8.2 Alpine Vegetation

The alpine belt in the Caucasus extends from 2,400–2,500 up to 2,900–3,000 m a.s.l. (Fig. 8.9) It is characterized by the dominance of short grass meadows; carpet-like alpine meadows, alternating with *Rhododendron caucasicum* thickets and rock-scree vegetation are also well developed here.

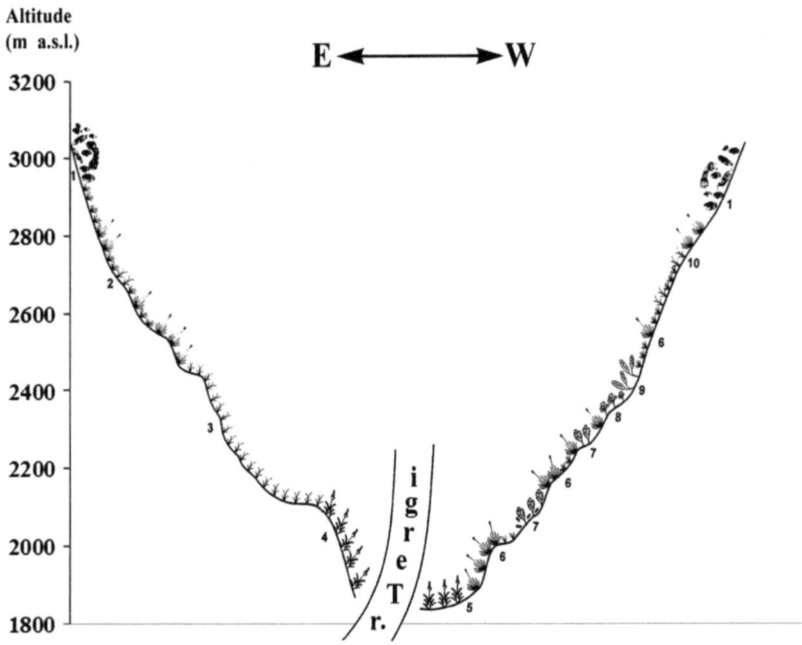

Fig. 8.9 Vegetation transect in Kazbegi region

8.2.1 Alpine Meadows

Alpine meadows occupy vast areas in Georgia. They are valuable natural grasslands. These communities mainly occur on mountain wetlands. Within the alpine meadows, it is possible to distinguish firm-bunch and sedge meadows, forb meadows and those dominated by tussock grasses and forbs. These groups of formations are typical for all geographical variants of the mountain-meadow vegetation of the Caucasus. The occurrence of local endemics is one of the floristic peculiarities of the above-mentioned communities (Bedoshvili 1985).

Information on the syntaxa of high-mountain vegetation (Svaneti region), including that of alpine belt is available in Table 8.10.

Festuceta variae, Festuceta supinae, Nardeta strictae, Cariceta tristis, Kobresieta capilliformis and K. persicae are the major formations of firm-bunch meadows.

Festuceta variae (*Festuca varia* subsp. *woronowii*). These communities are confined to steep slopes of southern exposure. Most frequently they occur in the eastern part of Kavkasioni.

Variegated fescue meadows are widespread in subalpine belt; they occupy areas, once populated by crook-stem forests, *Rhododendron* thickets and *Nardus* communities.

Nardeta. *Nardus* communities are very abundant throughout the Caucasus; they are restricted to the slopes of all exposures. Though due to the overgrazing, the secondary Nardus communities prevail here. Mixed communities of Nardeto-Festucetum variae

Table 8.10 The phytosociological classification of the vegetation of alpine belt of upper Svaneti (Dolukhanov et al. 1942)

Communities
1.1. *Sibbaldia parviflora* + *Carum meifolium*
2.1. *Sibbaldietum mixtoherbosum*
2.2. *Sibbaldia semiglabra* + *Geranium gymnocaulon*
2.3. *S. semiglabra* + *Festuca supina*
2.4. *S. semiglabra* + *Taraxacum stevenii*
2.5. *S. semiglabra* + *Ranunculus svaneticus*
2.6. *S. semiglabra* + *Pedicularis condensata*
2.7. *Sibbaldietum lichenoso–muscosum*
2.8. *Sibbaldia semiglabra* + *Campanula biebersteiniana*
2.9. *S. semiglabra* + *Ranunculus svaneticus* + *Gnaphalium supinum*
2.10. *S. semiglabra* + *Gnaphalium supinum*
2.11. *S. semiglabra* + *Carex micropodioides* + *Festuca supina*
3.1. *Ranunculus svanetici typicum*
3.2. *Ranunculus svaneticus* + *Gnaphalium supinum*
3.3. *R. svaneticus* + *Corydalis conorrhiza* + *Taraxacum stevenii*
3.4. *R. svaneticus* + *Pedicularis crassirostris*
4.1. *Caretum caucasici*
4.2. *Carum caucasicum* + *Trifolium ambiguum*
4.3. *C. caucasicum* + *Festuca supina*
5.1. *Minuartia colchica* + *Carum caucasicum*
5.2. *Minuartia aizoides* + *Campanula biebersteiniana*
5.3. *Plantago saxatilis* + *Festuca supina* + *Minuartia aizoides*
5.4. *Antennaria caucasica* + *Campanula saxifraga*
5.5. *Taraxacum stevenii* + *Gnaphalium supinum*
5.6. *Anthemis rudolphiana* + *Sibbaldia semiglabra*
5.7. *Festuca supina* + *Campanula saxifraga*
5.8. *Antennarietum caucasicae*
5.9. *Plantago saxatilis* + *Festuca supina*
5.10. *Antennaria caucasica* + *Alchemilla caucasica*
5.11. *Festuca supina* + *Chamaesciadium acaule* + *Plantago saxatilis*
6.1. *Alchemilletum caucasicae*
6.2. *Alchemilla caucasica* + *Antennaria caucasica* + *Festuca supina*
7.1. *Calamagrostis arundinacea* + *Betonica macrantha*
7.2. *Trollieto–Calamagrostidetum mixtoherbosum*
7.3. *Calamagrostidetum mixtoherbosum*
7.4. *Calamagrostis arundinacea* + *Anemonastrum fasciculatum*
8.1. *Deschampsieto–Calamagrostidetum mixtograminosum*
8.2. *Brometo–Calamagrostidetum mixtograminosum*
8.3. *Calamagrostis arundinacea* + *Trisetum pratense*
9.1. *Festuca djimilensis* + *Vaccinium myrtillus*
9.2. *Festuca djimilensis* + *Inula orientalis* + *Trollius ranunculinus*
10.1. *Trifolieto–Brometum mixtoherbosum*

(continued)

Table 8.10 (continued)

Communities
10.2. *Careto–Brometum mixtoherbosum*
11.1. *Trisetum pratense* + *Euphorbia abchasica* + *Polygonum carneum*
12.1. *Festuca djimilensis* + *Calamagrostis arundinacea* + *Geranium platypetalum* + *Vaccinium myrtillus*
12.2. *Calamagrostidetum arund. mixtograminoso–mixtoherbosum*
12.3. *Calamagrostidetum glaucae mixtograminoso–mixtoherbosum*
12.4. *Calamagrostis arundinacea* + *Festuca djimilensis* + *Anemonastrum fasciculatum* + *Betonica macrantha*
12.5. *Poetum longifoliae mixtoherbosum*
13.1. *Agrostis capillaris* + *Trifolium canescens*
14.1. *Trollius ranunculinus* + *Geranium gymnocaulon*
15.1. *Anemonastrum fasciculatum* + *Betonica macrantha*
16.1. *Alchemilla tredecimbola* + *Valeriana colchica*
16.2. *Alchemilletum tredecimlobae mixtoherbosum*
17.1. *Geranium typicum*
17.2. *Geranium gymnocaulon* + *Calamagrostis arundinacea*
18.1. *Geranium gymnocaulon* + *Carex aequivoca* + *Phleum alpinum* + *Anthoxanthum alpinum* + *Carum meifolium*
18.2. *Geranium gymnocaulon* + *Festuca supina* + *Senecio taraxacifolium* + *Aster caucasicus* + *Primula amoena*
18.3. *Triseto–Graminetum mixtoherbosum*
19.1. *Geranium gymnocaulon* + *Anemonastrum fasciculatum* + *Betonica macrantha*
19.2. *Geranium gymnocaulon mixtoherbosum*
19.3. *Geranium gymnocaulon* + *Trollius ranunculinus* + *Anemonastrum fasciculatum*
19.4. *Geranium gymnocaulon* + *Sibbaldia semiglabra*
19.5. *Geranium gymnocaulon* + *Campanula biebersteiniana*
19.6. *Geranium gymnocaulon* + *Trollius ranunculinus*
19.7. *Geranium gymnocaulon* + *Alchemilla sp.*
20.1. *Heracleetum (Heracleum aff. umbonatum Boiss.)*
20.2. *Ligusticetum alatae*
21.1. *Aconitum nasutum* + *Cephalaria gigantea* + *Heracleum aff. umbonatum*
21.2. *Senecio platyphyllus* + *Inula orientalis*
22.1. *Athyrium alpestris*
23.1. *Senecio platyphyllus* + *Rumex alpinus*
24.1. *Festucetum variae typicum*
25.1. *Festuca varia* + *Calamagrostis arundinacea*
25.2. *Festuca varia* + *Deschampsia flexuosa*
26.1. *Festuca varia* + *Alchemilla sp. div.*
26.2. *Festuca variae mixtoherbosum*
27.1. *Nardus stricta* + *Agrostis capillaris*
27.2. *Nardus stricta* + *Sibbaldia parviflora*
27.3. *Nardus stricta* + *Carum meifolium*
27.4. *Nardus stricta* + *Alchemilla sp.* + *Trifolium ambiguum*
27.5. *Nardus stricta* + *Plantago saxatilis*

(continued)

Table 8.10 (continued)

Communities
28.1. *Carex meinshauseniana* + *Anthemis rudolphiana*
28.2. *Caricetum typicum*
29.1. *Festuca supina* + *Carex meinshauseniana*
30.1. *Festuca ruprechtii* + *Geranium renardii*
31.1. *Deschampsia flexuosa* + *Geranium renardii*
32.1. *Deschampsia flexuosa* + *Festuca supina* + *Anthemis rudolphiana*
33.1. *Deschampsia flexuosa* + *Anthemis rudolphiana* + *Campanula saxifraga*
34.1. *Kobresieta schoenoides*
35.1. *Carex dacica* + *Primula auriculata*
35.2. *Carex dacica* + *Primula auriculata* + *Trifolium ambiguum*
36.1. *Alchemilla tredecymloba* + *Deschampsia cespitosa*
37.1. *Rhododendronetum typicum*
38.1. *Rhododendron caucasicum* + *Chamaenerium angustifolium*
38.2. *Rhododendron caucasicum* + *Geranium gymnocaulon*
38.3. *Rhododendron caucasicum* + *Athirium alpestre*
39.1. *Rhododendron caucasicum* + *Vaccinium myrtillus* + *Deschampsia flexuosa*

can be found in the central and eastern parts of the Greater Caucasus. Under the influence of pasturing hygrophytic *Nardus* communities have developed on swampy depressions of alpine belt.

Cariceta tristis. These coenoses are very abundant in the high-mountains of the Caucasus. Usually they inhibit prominent slopes exposed to winter winds. *Carex tristis* participates in the formation of sedge-fescue association *(Festuca supina-Carex tristis)*. The most typical communities are such, as: Cariceto-Alchemilleta *(Alchemilla caucasica)* and Cariceto-Kobresieta *(Kobresia capilliformis)*.

Kobresieta capilliformis. These communities like those of *K. persica* are found chiefly in the central and eastern parts of Kavkasioni and the southern upland of Georgia. They cover mountain ridges and prominent slopes. Coenoses, with Middle Asian-Himalayan *Kobresia capilliformis*, are very common in Georgia; the above-mentioned communities occur on marl-like shales and limestones of the Central Caucasus. Within variegated fescue meadows Kobresieta schoenoides occur on stony localities (Table 8.11).

Festuceta djimilensis. It belongs to the tussock grass-forb meadows, which have penetrated into the alpine belt from subalpine zone; Festuceta djimilensis is typical for both cobble and wet substratum (mostly in the western part of Kavkasioni).

Bromopsieta variegati, which enter the same group, are restricted to the southern slopes of alpine belt. In the high-mountains of Colchis, dry meadows occur on limestones; these meadows are formed by *Sesleria anatolica*, a grass with creeping rhizome.

Geranieta (*Geranium gymnocaulon*). These communities of forb meadows of alpine belt, are distributed in the western part of Kavkasioni. According to

Table 8.11 Kobresieta in Kazbegi

Height (m s. m.)	2,150
Exposition	N
Slope (°)	5
Cover (%)	90
Kobresia capilliformis	5
Kobresia persica	4
Carex buschiorum	2
Trifolium ambiguum	2
Trifolium repens	2
Trifolium fontanum	1
Trifolium canescens	1
Agrostis planifolia	1
Helictotrichon asiaticus	2
Bromopsis variegata	1
Lotus caucasicus	2
Polygonum viviparum	2
Antennaria caucasica	2
Campanula biebersteiniana	2
Leontodon danubialis	+
Taraxacum confusum	2
Plantago caucasica	2
Minuartia circassica	2
Minuartia oreina	2
Cerastium purpurascens	2
Draba hispida	1
Draba repens	1
Cirsium pugnax	+
Gnaphalium supinum	+
Thesium alpinum	+

Dolukhanov (1946), Onipchenko and Blnnikov 1994 *Geranium* communities extend from the low-alpine subzone to the upper limits of dense vegetation; these meadows merge into the subnival belt. They occur on mountain-meadow skeletal soils.

8.2.2 Carpet-Like Alpine Meadows

Carpet-like alpine meadows are a diverse group of formations. They occur in the upper part of the alpine belt. These coenoses are to be found among large stones.

The term "carpet-like alpine meadows" proposed by Grebenshikov (1965) coincides with "Dicotylen-Teppiche" of Ellenberg (1996). Alpine carpet-like meadows are composed of the following low rosetted and caespitose plants: *Campanula biebersteiniana, Carum caucasicum, Veronica gentianoides,*

Gnaphalium supinum, Pedicularis crassirostris, P. armena, Poa alpina, Taraxacum stevenii, T. porphyranthum, Sibbaldia semiglabra etc.

These communities include such grasses as *Poa alpina, Phleum alpinum, Festuca supina,* etc.

Carpet-like alpine meadows differ from true meadows by their floristic composition and ecological peculiarities. Due to the lasting snow cover, they are characterized by short vegetative period. Herbage of the above-mentioned meadows does not exceed 3–4 cm. Eighty percentage of phytomass is accumulated in a 0–1 cm layer. As noted above, the carpet's sod belongs to the forb-type sod-forming. The typical carpet's sod has a two-layer structure. The main mass of the individual sod is concentrated in the upper layer (5–6 cm). The lower layer consists of separate deep roots (20–25 cm). Alpine carpet-like meadows are distinguished by large and brightly coloured flowers and inflorescences.

Some authors suggest considering alpine carpet-like meadows as Nanoherbeta, which belong to eumesophytic meadows.

Besides the climax carpet-like alpine meadows, secondary carpets, enriched by meadow elements (especially *Alchemilla* species), occupy considerable areas. Development of these secondary carpets is caused by overgrazing.

Sibbaldia semiglabra communities as well as those of *S. parviflora* are of particular interest. Pure *Sibbaldia* communities are very poor floristically.

These secondary communities occupy the most exploited pastures. Endemic carpet-like alpine meadows with *Ranunculus helenae* occur on limestone substratum in Colchis. Moist sites are occupied by carpets of *Carum caucasicum.* Pure stands of the above-mentioned plant are quite rare; usually it associates with *Taraxacum stevenii, Campanula biebersteiniana, Plantago saxatilis, Minuartia aizoides,* etc.

Carpet-like communities which are developed around the snow spots and at the edges of glaciers, include the following species: *Ranunculus oreophilus var. pumilus, R. baidarae, Primula algida, Gentiana djimilensis (= G. pyrenaica), G. angulosa, G. nivalis, Minuartia aizoides* and *Cerastium cerastoides.*

Of the plants growing at the banks of brooks, the following may be mentioned: *Pedicularis crassirostris, Primula auriculata, P. nordmanniana, Poa alpina,* etc.

8.2.3 Alpine Shrub

Steep northern slopes of alpine belt are inhibited by *Rhododendron caucasicum* thickets (dekiani). *Rhododendron* communities occur even at an altitude of 1,700–1,800 m (Table 8.12). According to Kolakovsky (1961), at greater altitudes *Rh. caucasicum* is restricted to cirques and valleys, which are characterized by high humidity and profound snow, protecting it from frost. The same scholar, attributes the present-day distribution of *Rh. caucasicum* in alpine belt to the lowering of upper forest margins, to which it has been confined before.

Rhododendron thickets occur on brownish-cinnamon bog soils, which are very rich in humus (usually skeletal) (Fig. 8.10).

Table 8.12 *Rhododendreta* i Kazbnegi

Height (m s. m.)	2,450
Exposition	N
Slope (°)	25
Cover (%)	100
Rhododendron caucasicum	5
Vaccinium myrtillus	3
Vaccinium vitis-idaea	2
Calamagrostis arundinacea	2
Pyrola minor	1
Agrostis planifolia	2
Anemonastrum fasciculatum	1
Geranium ibericum	1
Empetrum caucasicum	2
Luzula spicata	+
Carex tristis	+
Nardus stricta	+

Subshrub communities, formed by *Dryas caucasica* (Table 8.13) arise much interest. They favour cold stony slopes of northern exposure. In Colchis, *Dryas* communities are restricted to limestone rocks, rocky mountain ridges and stony slopes of northern exposure (Kolakovsky 1961).

Distribution area of elfin woods of *Juniperus hemisphaerica* and *J. sabina* extends throughout Kavkasioni. In Colchis, the fragments of *Juniperus* communities (Fig. 8.11) rarely intermingle with *Rhododendron* thickets. Juniper coenoses with *Woronowia speciosa (= Geum speciosum)* occupy limestone mountains (Kolakovsky 1961).

Chionophilic communities of *Daphne glomerata* are very characteristic of Kavkasioni. Coenoses of *D. albowiana* and *D. woronowii* are abundant in Colchis. It should be noted, that *D. woronowii* communities occur on limestones.

8.2.4 Alpine Saxicolous Plants

Of the saxicolous plants of alpine belt the following should be outlined: *Campanula mirabilis, C. dzaaku, Arenaria lychnidea, Draba bryoides, Albowiodoxa elegans, Symphyandra armena,* etc. The screes are occupied by *Heracleum calcareum, Campanula schistosa, Athyrium alpestre, Anthemis sosnovskyana, Barbarea minor,* etc.

8.2.5 Ecological Characteristics of Alpine Communities

Ecological characteristics of Nardeta strictae and Cariceta tristis are presented bellow. These communities are typical representatives of alpine vegetation of the

Fig. 8.10 *Rhododendron caucasicum* & *Betula litwinowii* at treeline, Kazbegi, Central Caucasus (Photo O. Abdaladze)

Caucasus. They are confined to the central part of the Greater Caucasus (Kazbegi region, 2,650 m).

8.2.5.1 Soils

Mother rocks of this area are of volcanic origin. Soils, formed by volcanic ash, are becoming andosols or Andept (the term proposed by North American pedologists). Soils populated by the afore-mentioned communities have A-B-C profile. Transition to A–C soils can be observed at greater altitudes. Soil pH of A layer is very acid, sandy loam and pierced through with roots. It appears that water content in phytocoenoses accounts for the changes of soil color; thus, when lacking water, the colour of soil becomes lighter.

Table 8.13 *Dryeta* in Kazbegi

Height (m s. m.)	2,050
Exposition	NW
Slope (°)	50
Cover (%)	100
Deschampsia flexuosa	3
Anthoxanthum alpinum	3
Carex meinshauseniana	3
Agrostis planifolia	2
Bromopsis variegata	2
Helictotrichon asiaticus	2
Nardus stricta	1
Poa alpina	1
Festuca supina	2
Daphne glomerata	2
Vaccinium vitis-idaea	1
Dryas caucasica	4
Carum caucasicum	3
Minuartia circassica	1
Trifolium ambiguum	2
Trifolium trichocephalum	1
Vicia alpestris	1
Primula amoena	1
Alchemilla sericata	1
Ranunculus caucasicus	1
Viola somchetica	1
Leontodon hispidus	1
Anemonastrum fasciculatum	1
Carum alpinum	1
Gentianella caucasica	1
Polygonum viviparum	2
Campanula collina	1
Betonica macrantha	1
Plantago caucasica	2
Selaginella selaginoides	1
Selaginella helvetica	1
Leontodon danubialis	1
Polygala alpicola	1
Polygonum carneum	1
Cirsium obvallatum	1
Ranunculus acutilobus	1
Cruciata laevipes	1
Pyrethrum roseum	1
Ranunculus oreophilus	2
Cirsium simplex	+

(continued)

Table 8.13 (continued)

Parnassia palustris	+
Silene ruprechtii	+
Rhinanthus minor	+
Cirsium pugnax	+
Taraxacum confusum	+
Cerastium purpurascens	+
Minuartia imbricata	+
Scabiosa caucasica	+
Inula orientalis	+

The content of organic water (17–20 %) is rather high in A layer of both phytocoenoses. However, at an altitude of 1,970 m, the same index is 14 %.

8.2.5.2 Temperature Regime

On sunny days the herbages of both phytocoenoses undergo overheating. In Cariceta the lower layers and soil surface represent the "active surface" of conductive heat exchange. For example, the mean diurnal temperature of soil surface is 9.4 °C higher than that of the air at the height of 2 m, and it is 10.5 °C higher – at 1 m.

When studying microclimate and energy balance, we have applied the methods of Cernusca (1976).

Maximum temperature of soil surface is 29 °C, more than that of the air at the height of 2 m.

In Nardeta the overheating of herbage is less than in Cariceta. Solar radiation is allocated over more expansive biomass, which accounts for less overheating. Maximum overheating is observed in the lower 3 cm layer of phytocoenoses. In this layer of herbage, the average diurnal temperature is 6.7 °C higher and maximum temperature is 19.4 °C higher than that of air at the altitude of 2 m. In Nardeta on soil surface the temperature reaches its maximum meaning 2 h later, than at 3 cm higher above soil. This must be provided by a rather thick layer of detritus over the soil, which is characterized by low heat conductivity.

The temperature of leaves is very important for photosynthesis, respiration and transpiration. By the peculiarities of temperature of leaves, both phytocoenoses are of two-layer structure: in lower warm layers of herbage (Cariceta 0–1 cm, Nardeta 0–5 cm) the temperature of leaves is below the air temperature. In Nardeta at the height of 0.5 cm leaves are cooled off on average 1.3 °C during the day, and in Nardeta at the height of 4 cm – on 2.6 °C. Under the influence of wind, intensive convective heat exchange takes place in the upper layer of phytocoenoses. The temperature of leaves here almost equals with that of the air. On the average, the temperature of leaves in Nardeta (height −3 cm) exceed that of the air just by 0.1 °C daily.

Air humidity, especially water vapour pressure gradients between leaves and their environment are very essential factors for many ecosystems.

Fig. 8.11 *Juniperus communis* (Photo O. Abdaladze) in Eastern Central Caucasus

The lack of water vapour pressure at different height levels from the surface of soil distinguishes two layers of phytocoenoses structure. In Cariceta tristis the layer above 1 cm is strongly influenced by wind and dry air; during the one hour (between 1.00 and 2.00 a.m.) the water vapour deficit amounts to 37.6 mbar at the height of 3 cm, whereas at 1 cm above the surface of soil water pressure deficit is only 17.2 mbar.

8.2.5.3 Solar Radiation Regime

On sunny days, daily global radiation equals to 562 cal. Cm^{-2}. In Cariceta tristis 21.9 % of this amount is reflected by leaves, while in Nardeta only 19.2 %. The value of albedo of Cariceta tristis coincides with that of Cariceta curvulae; these communities occur in Hohe Tauern (Cernusca 1976). In Cariceta 358 cal. Cm^{-2} is absorbed in the form of radiation balance, while in Nardeta 338 cal. cm^{-2}.

For Cariceta curvulae in the Hohe Tauern Mts. (Austrian Alps) the radiation balance makes up 66 % of global radiation.

In good weather, 44 % of absorbed solar energy (radiation balance) in *Cariceta* is spent on evapotranspiration, 40.7 % on air heating and 15.3 % on soil heat flow, correspondingly, in Nardeta it is 62.5 %, 28.7 % and 8.8 %.

Consequently, the soil heat flow in Cariceta is twice as much as in Nardeta. It can be accounted for by the fact, that a large amount of radiation is being absorbed in the lowest layers of phytocoenoses (0–0.5 cm) and in the soil surface. Whereas, in Nardeta radiation is absorbed in the upper layers. Besides, Nardeta has a rather dense layer of detritus, which retains the soil heat flow.

Results of the analysis confirm the conclusions based on the studies of dwarf shrub heaths in Patscherkofel (Tyrol) (Cernusca 1976) and alpine heaths in Hohe Tauern mts. Obviously, the structure of phytocoenose (orientation of leaves, shape, height of plants) and exposure have a great impact on the microclimate and energy balance of high-mountain ecosystems. Both of the studied phytocoenoses are of two-layer structure. The upper dry and relatively cool layers are more influenced by wind. The lower warm and moist layers are characterized by negligible wind speed. At the same time, we have found out that the structure of phytocoenose, microclimate and energy balance of Cariceta tristis in the Caucasus is similar to those of Caricetum curvulae in Hohe Tauern.

8.2.5.4 Water Potential

Alpine plants of the Caucasus are characterized by higher value of stomatal conductance than those of Alps (Körner 2003). Absence of stomata closing tendency, which is more pronounced in Nardeta, indicates at good water supply of both phytocoenoses. The values of stomatal conductance and water potential of leaves of *Carex tristis* seems to change in the opposite direction.

Extremely low values of maximum diffusion resistance speak of sufficient water supply and moist growth conditions of the plants, which occur in the alpine belt of Georgia.

8.3 Subnival Vegetation

The subnival belt is well represented in the Caucasus. It is situated between the belt of alpine meadows and that of firm fields at the altitude from 2,900–3,000 up to 3,500–3,700 m a.s.l.

In the Greater Caucasus, lower limit of the above-mentioned belt varies in altitude between 2,900 and 3,100 m, increasing from west to east. Due to the low position of eternal snow line (about 2,800 m) in the western part of the Greater Caucasus, boundaries of subnival belt, represented here fragmentally, are not subject to classification.

8.3.1 Life Conditions in Subnival Zone

Subnival belt of the Caucasus is characterized by moist climate with short cold summer and long severe winter. On the basis of the data of observations by the meteorological station, situated at 3,656 m a.s.l. (Kazbegi), the subnival belt can be characterized by the following figures: mean annual temperature 6.1 °C, mean temperature of January −15 °C, mean temperature of the warmest months (July–August) –10 °C; extreme maximum +16 °C; extreme minimum −42 °C, annual amount of precipitation is 1,074 mm, mean annual wind speed 6.4 m/s.

In the extreme high-mountain conditions life activity of plants is mainly governed by the temperature of boundary air layer.

According to the represented tautochrones (Fig. 8.12) high air temperature during the daytime (from 12.00–13.00 to 16.00–17.00) is observed at 0–10 cm above the soil surface. Temperature of the earth surface runs to high values, but at the depth of 10–20 cm it remains relatively low during the whole day. In the morning, the soil surface temperature is almost always negative.

The role of temperature as a limiting factor becomes more important; excessive solar radiation raises the temperature of soil surface and of boundary air layer, daily temperature fluctuations and evapotranspiration also increase. The role of wind as an ecological factor should be particularly emphasized, as the character of distribution of plants and communities in subnival and nival belts in many respects is dependent on the direction and strength of wind.

According to Larcher (1980; Larcher et al. 2010), Körner (2003), wind direction in high-mountains determines great differences in habitats, water and temperature relations of plants and soil. In respect of the observations produced by the same scholar, rocks, ridges and mountain-sides exposed to wind are less intensively warmed up by sun rays than the elongated cavities and ditches, protected from wind. The efficiency of CO_2 assimilation amounts with the altitude.

The pattern of plant microgrouping distribution, temperature and water relations of habitats, the phenological rhythm type of plants in subnival and especially nival belts are also predetermined by snow cover. In high-mountains some species of plants (chionophobes) are well adapted to low temperatures and others (chionophiles) have low frost resistance (Larcher 1977, 1987; Sakai and Larcher 1987).

One of the clearly defined characteristic properties of high-mountain biomes of diverse globe zones, including subnival and nival belts of the Caucasus, is a great variety of ecological niches within even small territories. Metabolism and plant growth in the extreme conditions of high-mountains are mainly governed by characteristic properties of microclimate of these niches.

Another unfavourable environmental factor affecting the growth of plants at high altitudes, is the ultraviolet radiation. However, due to the high concentration of carotinoids and flavonoids, these plants are well adapted to this factor (Caldwell 1968, 1989).

Physical and geographical conditions of the subnival belt of the Caucasus are particularly extreme, but due to life strategy adaptations, for some plants, typical for this belt, such living conditions are quite optimal.

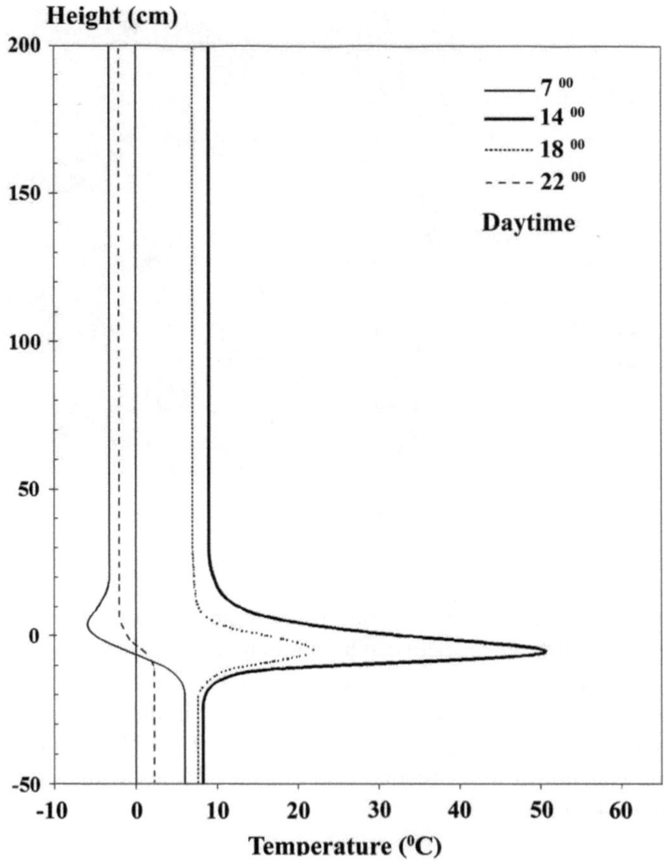

Fig. 8.12 Temperature tautochrones, Elbrus, 3.500 m.a.s.l.

In connection with the absence of terrains with more or less mature soil, almost all the plants of subnival ecosystems are rock and talus plants. Very high percentage of endemic, including also narrow-endemic species are prominent. Some of them fall into mono- or oligotype genera (*Pseudobetckea, Symphyoloma, Pseudovesicaria, Didymophysa, Trigonocaryum, Eunomia, Vavilovia*).

Severe climatic conditions dictate strict reglementation of population viability of every species, which has its own life strategy, habitat, favourable to development of individuals in every population, often scattered at considerable distances create quite often favourable conditions for the development of another one. In this connection, competition between populations of different species often seems to be masked or even eliminated by positive effect of different plant coexistence.

Fig. 8.13 *Cerastium kazbek*, the Central Caucasus (Photo Z. Kikodze)

Besides the peculiarities of climate, a lithologic composition of waste mantle and the degree of differences between high mountain conditions and the conditions similar to ones in edaphic and climatic respects, have a great impact on floristic composition of species, closely related to either rock, in the majority of cases have narrow area, whereas among the species indifferent to lithologic base of waste mantle, one can find fewer endemic species; the species with wide distribution predominate (Dolukhanov 1969).

A large amount of endemic species, which are typical representatives of subnival flora and even endemic genera, present enough grounds to suggest the existence of a suitable floristic complex, within the region, with analogues of high-mountain landscapes, in the geological past, far beyond the Pleistocene (Kharadze 1965).

8.3.2 Flora

Extreme environmental conditions appear in the subnival zone of the Caucasus. Nevertheless, more than 300 species occur here, among them plants which are typical for alpine and sometimes for subalpine belts (Table 8.14). Only 109 species are characteristic for the subnival zone (Kharadze 1965). Location of upper boundary of flowering plants distribution depends on the level of continental climate and on highness of mountain. For example in the highest region of the Greater Caucasus (Elbrus, Shkhara, Kazbegi) flowering plants reach 3,900–4,500 m a.s.l., while in continental Dagestan (East Greater Caucasus) (Prima 1974) and in mountain Aragatz (Lesser Caucasus) (Voskanyan 1976) they occur correspondingly at 3,300–4,000 m. At 4,000 m (Central Caucasus) appears *Cerastium kazbek* (Fig. 8.13) and at 4,000 m (Mt. Aragatz, Lesser Caucasus) *Draba araratica*. The following typical subnival

Table 8.14 List of plants growing in the subnival belt of Georgia

Apiaceae
Chaerophyllum humile
Symphyoloma graveolens
Asteraceae
Aetheopappus caucasicus
Antennaria caucasica
Anthemis iberica
Anthemis sosnovskyana
Erigeron uniflorus
Gnaphalium supinum
Jurinea filicifolia
Jurinella moschus
Jurinella subacaulis
Podospermum alpigenum
Podospermum meyeri
Senecio karjaginii
Senecio sosnovskyi
Senecio taraxacifolius
Taraxacum porphyranthum
Taraxacum stevenii
Tripleurospermum caucasicum
Tripleurospermum subnivale
Boraginaceae
Myosotis alpestris
Brassicaceae
Arabis kazbegi
Draba bryoides
Draba siliquosa
Draba supranivalis
Eunomia rotundifolia
Pseudovesicaria digitata
Thlaspi pumilum
Campanulaceae
Campanula saxifraga
Caryophyllaceae
Arenaria lychnidea
Cerastium cerastoides (=Dichodon cerastoides)
Cerastium kasbek
Cerastium multiflorum
Cerastium polymorphum
Cerastium undulatifolium
Minuartia imbricata
Minuartia inamoena
Minuartia ruprechtiana

(continued)

Table 8.14 (continued)

Minuartia trautvetteriana
Silene humilis
Silene lychnidea
Silene marcowiczii
Crassulaceae
Sedum involucratum
Sedum tenellum
Sedum stevenianum
Cyperaceae
Carex tristis
Fabaceae
Trifolium polyphyllum
Fumariaceae
Corydalis alpestris
Corydalis emanuelii
Murbeckiella huetii
Iridaceae
Crocus scharojanii
Juncaceae
Luzula spicata
Lamiaceae
Lamium tomentosum
Nepeta supina
Thymus nummularius
Ziziphora puschkinii
Ziziphora subnivalis
Liliaceae
Lloydia serotina
Poaceae
Alopecurus dasyanthus
Alopecurus glacialis
Alopecurus vaginatus
Colpodium versicolor
Festuca supina
Poa alpina
Poa caucasica
Trisetum buschianum
Trisetum spicatum
Primulaceae
Androsace albana
Primula bayernii
Ranunculaceae
Delphinium caucasicum

(continued)

Table 8.14 (continued)

Rosaceae
Alchemilla chlorosericea
Alchemilla sericea
Alchemilla caucasica
Potentilla gelida
Sibbaldia semiglabra
Rubiaceae
Galium coronatum
Saxifragaceae
Saxifraga exarata
Saxifraga flagellaris
Saxifraga moschata
Saxifraga ruprechtiana
Saxifraga scleropoda var.nivalis
Saxifraga sibirica
Saxifraga subverticillata
Scrophulariaceae
Scrophularia minima
Veronica minuta
Valerianaceae
Valeriana alpestris
Valeriana saxicola
Pseudobetckea caucasica
Violaceae
Viola minuta

species are found at 3,300–3,800 m a.s.l.: *Saxifraga moschata, S. exarata, S. flagellaris, S. sibirica, Tripleurospermum subnivale, Colpodium versicolor, Alopecurus dasyanthus, Draba supranivalis, Veronica minuta, V. telephiifolia, Senecio karjagini, Cerastium pseudokasbek, Pseudovesicaria digitata.*

A number of endemic species and genera occur in the subnival zone: monotypic genera – *Pseudovesicaria* (Brassicaceae), *Symphyoloma* (Apiaceae), *Pseudobetckea* (Valerianaceae).

From Caucasian-Asia Anterior–South-West Asian genera there are found oligotypic *Coluteocarpus*, *Didymophysa* and *Eunomia* (Brassicaceae), *Vavilovia* (Fabaceae). They are mainly distributed in high mountains of Lesser Caucasus and Asia Anterior–South-West Asia. Almost missing are Cyperaceae and shrubby plants. The same phenomenon is observed in polar deserts (Tikhomirov 1963; Aleksandrova 1983) the composition of the flora with respect to the soils is different. On the volcanic rocks of Keli Elevation has been established flora somehow different from one of mergel slates. High percentage of Caucasian endemics are connected with mergel slates (Dolukhanov 1969). Especially are distinguished limestone-built rocks of Western Caucasus where a number of local endemics appear (Nakhutsrishvili and Gagnidze 1999).

Fig. 8.14 *Delphinium caucasicum* (Photo O. Abdaladze)

8.3.3 Vegetation

Under subnival vegetation we consider the vegetation of a certain altitudinal zone placed between alpine low herbaceous meadows and nerve plain. Distribution of contagious vegetation is limited and open groups dominate mainly.

The following types of vegetation of the subnival zone are noticed.

1. Open groups are formed by one population or by few species of flowering plants, which are not contiguous to each other either by overground or by underground parts. Only biotope and environmental conditions are common for them.
2. Nanocoenoses or ultramicrocoenoses are dominated by individuals of vascular plants sprinking with turfs of certain plants or lichens. Plants contact with underground and overground parts. Nanocoenoses are spread in patches.
3. Fragments of turfy alpine meadows established by grasses and alpine carpets formed by various herbaceous plants. Fragments of alpine meadows have broad ecological amplitude, whereas "carpets" are found only at the snow edges. Overground and underground relationship between plants is well presented.

Large amount of mosses and lichens occur in all types of vegetation.

Mosses: *Bryum* sp., *Dicranum elongatum, Dicranoweisia crispula, Pohlia elongata, Pogonatum nanum, Tortulla tortuosa, Tortulla muralis* etc.

Lichens: *Thamnolia vermicularis, Hypogimnia encausta, Cetraria islandica, C. nivalis, Parmelia vagans, Umbilicaria virginis, Placolecanora melanophthalma, P. rubina, P. murilis, Stereocaulon alpinum, Caloplaca elegans, Umbilicaria cylindrica, Lecidea atrobrunnea, Rhizocarpon geographicum,* etc.

Species from open groups are presented in all biotopes of subnival zone with the exception of snow edges. For example on temporarily moving screes there occur: *Cerastium kasbek, Delphinium caucasicum* (Fig. 8.14); on weekly moving: *Veronica minuta, V. telephiifolia, Scrophularia minima* (Fig. 8.15), *Lamium tomentosum* (Fig. 8.16); – on fixed screes: *Aetheopappus caucasicus, Symphyoloma graveolens,*

Fig. 8.15 *Scrophularia minima* (Photo O. Abdaladze)

Fig. 8.16 *Lamium tomentosum* (Photo O. Abdaladze)

Jurinella subacaulis, J. moschus, Minuartia inamoena, Viola minuta (Fig. 8.17), *Eunomia rotundifolia* (Fig. 8.18); – on rocks: *Primula bayernii, Draba bryoides, Saxifraga moschata, S. exarata*. All these plants are chionophobes or hemichionophobes and are different from each other by living form. Espalier plants are: *Veronica minuta, V. telephiifolia*; – acaulescent rosette plants: *Scrophularia minima, Symphyoloma graveolens, Aetheopappus caucasicus*; – cushions: *Minuartia inamoena, M. trautvetteriana, Saxifraga moschata, S. exarata, Draba bryoides*.

Fig. 8.17 *Viola minuta* (Photo O. Abdaladze)

Fig. 8.18 *Eunomia rotundifolia* (Photo O. Abdaladze)

Almost all above-mentioned plants are typical representatives of the subnival zone and are well adapted to the extreme environmental conditions.

Nanocoenoses are spread in patches through the whole subnival zone of the Caucasus. Their dimensions sometimes are 15–20 cm^2. Number of plant rarely exceeds 4–5 species (Table 8.15). Nevertheless different aged populations are found there: sprouts, juvenils, immature and generative individuals with the whole seasonal cycle (Nakhutsrishvili 1974; Nakhutsrishvili and Gamtsemlidze 1984). Subnival and alpine species are present in all nanocoenoses which give possibility to alpine plants for successful survival and expansion of area.

Table 8.15 Nanocoenoses of the subnival belt (the Mamisoni pass in the central Caucasus)

Screes of southern slopes of 5–7°
Potentilla crantzii – Poa alpina – Symphyoloma graveolens
Poa alpina – Minuartia aizoides – Alchemilla sericea – Anthemis sosnovskyana
Alchemilla sericea – Poa alpina – Anthemis sosnowskyana
Minuartia inamoena – Poa alpina – Symphyoloma graveolens – Anthemis sosnovskyana – Taraxacum stevenii – Veronica gentianoides
Minuartia aizoides – Poa alpina – Symphyoloma graveolens
Arenaria lychnidea – Anthemis sosnovskyana
Minuartia inamoena – Campanula saxifraga – Potentilla gelida
Minuartia inamoena – Minuartia aizoides – Saxifraga exarata – Poa alpina
Minuartia inamoena – Saxifraga exarata – Symphyoloma graveolens – Anthemis sosnovskyana – Poa alpina – Taraxacum stevenii
Arenaria lychnidea – Campanula biebersteiniana – Saxifraga exarata – Taraxacum stevenii
Scree biotope of western exposure (10°) at 3,200 m
Jurinella subacaulis – Saxifraga sibirica
Alopecurus glacialis – Tripleurospermum subnivale – Erigeron uniflorus
Alopecurus glacialis – Tripleurospermum subnivale
Alopecurus glacialis – Sedum tenellum
Poa alpina – Tripleurospermum subnivale – Draba supranivalis
Alopecurus glacialis – Saxifraga moschata
Eunomia rotundifolia – turf of mosses and lichenes
Mergel slates of western exposure at 3,300 m
Saxifraga moschata – Tripleurospermum subnivale
3,350 m Stony scree biotope (Mt. Elbrus)
Potentilla gelida – Carum caucasicum
Potentilla gelida – Minuartia trautvetteriana – Saxifraga flagellaris – Veronica minuta
Minuartia trautvetteriana – Veronica telephiifolia
Tripleurospermum subnivale – Carum caucasicum – Cerastium cerastoides – Veronica telephiifolia – Senecio karjaginii
Tripleurospermum subnivale – Veronica telephiifolia – Senecio karjaginii – Saxifraga flagellaris – Cerastium cerastoides – Anthoxanthum alpinum
Potentilla gelida – Minuartia trautvetteriana – Veronica telephiifolia – Saxifraga flagellaris

Analysis of spatial relationships among species in the subnival zone of the Central Caucasus in 300 vegetation patches showed that 144 of the studied patches contained only one species, whereas the other 156 contained 2.36 ± 1.31 species, with species numbers distributed as follows: 59 patches with two species, 41 with three, 39 with four and 17 with five species.

In the multi-species patches, the 22 most frequent species were examined and 46 statistically significant species associations, 36 positive and 10 negative, were found. Ten of these 22 species were typical subnival plants very rarely occurring at lower altitudes. The other 12 species are "invaders" as they have broader altitudinal ranges of occurrence and are common in alpine and even in the subalpine zones. Contrary to the typical subnival species the invaders were found exclusively

Table 8.16 Basic nanocoenoses of the subnival belt of the Mamisoni pass bu the altitudinal gradient

3,000 m
Alchemilla sericea – Poa alpina – Sibbaldia semiglabra
Saxifraga moschata – Alchemilla sericea – Minuartia aizoides
Festuca supina – Campanula biebersteiniana – Carum caucasicum – Alchemilla sericea
3,200 m
Alopecurus dasyanthus – Saxifraga moschata – Minuartia inamoena
Tripleurospermum subnivale – Scrophularia minima – Delphinium caucasicum
Alopecurus dasyanthus – Tripleurospermum subnivale – Erigeron uniflorus
Alopecurus dasyanthus – Sedum tenellum
Taraxacum stevenii – Campanula biebersteiniana – Carum caucasicum
3,300 m
Saxifraga moschata – Tripleurospermum subnivale – Alopecurus dasyanthus
Alopecurus glacialis – Cerastium polymorphum
Erigeron uniflorus – Lamium tomentosum – Potentilla gelida
3,400 m
Delphinium caucasicum – Saxifraga moschata – Lamium tomentosum
Tripleurospermum subnivale – Colpodium versicolor – Ziziphora puschkinii
Eunomia rotundifolia – Tripleurospermum subnivale
Saxifraga moschata – Cerastium cerastoides
3,500 m
Saxifraga moschata – Eunomia rotundifolia – Senecio karjaginii
3,600 m
Saxifraga moschata – Cerastium kasbek

in the patches with more than one species. Invader species were significantly more associated with subnival species while they were significantly less associated with other species than expected by chance. An interpretation is that typical subnival species nurse plants from lower altitudes and facilitate their invasion to more adverse subnival environments (Kikvidze and Nakhutsrishvili 1998).

Nanocoenoses are presented almost in all biotopes of subnival zone especially in the schistous substrate.

They often reach great altitudes, for example in Mamisoni Pass nanocoenoses decrease only from 3,600 m a.s.l. (Table 8.16).

The floristic composition of nanocoenoses occurring on the same relief is very alike.

To assess the homogeneity of vegetation in horizontal structure was studied by the method of Raunkiaer (Braun-Blanquet 1964). Three sections of Mamisoni Pass and one section placed at the foot of Mt. Elbrus were explored. Was used circle (diameter 35.6 cm) restricting circle 1/100 m^2 of area 50 times. In brief, this method states that the vegetation of any surface is floristically homogeneous when the classes of higher frequency in the diagrams contain more number of species and the classes of lower frequency, accordingly, contain less number of species and vice versa.

The first section is located at 2,950 m on the western slope of 5–7°. General cover is 30–40 %. From typical subnivale species here occur: *Aetheopappus caucasicus, Symphyoloma graveolens, Jurinella subacaulis* etc. Mainly are dominated species from alpine zone. Homogeneity of vegetation is low. The following plants have the greatest frequency: *Minuartia oreina, Veronica gentianoides, Anthemis sosnovskyana.*

Second section is situated on southern slope of 5–6° at 3,000 m. Cover is 30–40 %. Here appear subnival species: *Aetheopappus caucasicus, Silene marcowiczii, Symphyoloma graveolens.* However, still dominate species from alpine zone. Homogeneity is low as well. Greatest frequency is observed for *Minuartia oreina* and for *Aetheopappus caucasicus* and *Campanula saxifraga.*

The third section is placed on western slope of 5–7° at 3,100 m. Cover of vegetation is 20–30 %.

There occur subnival species: *Alchemilla sericea, Saxifraga moschata, Symphyoloma graveolens.* Nevertheless, presence of alpine plants prevail over subnival ones. Homogeneity is low.

Fourth section is located at the foot of mt. Elbrus (3,350 m). Mesorelief is plane, strongly stoned. With the subnivale species – *Minuartia trautvetteriana, Saxifraga flagellaris, Veronica minuta* – there appear typical alpine plants: *Carum caucasicum, Poa alpina, Cerastium cerastoides.* The following species have the high frequency: *Carum* caucasicum, *Veronica minuta, Potentilla gelida, Minuartia trautvetteriana, Saxifraga flagellaris.* Consequently, a little more homogeneity is characteristic for subnival zone especially on the transitional places from alpine and subnival belt.

Fragments of turfy meadows and alpine carpets are well developed in subnival belt. Ellenberg (1996) calls these fragments "Rasenstъcke" and he considers that they are typical primary alpine meadows. Reisigl and Pitschmann (1958) call the above-mentioned fragments "Pionierrasen" and the zone of their distribution "Pionierrasenstufe". Ellenberg accounts them as a formed and stabile climax association and refuses their farther development because of strongly restricted habitat. Composition of species very often remains to be stabile for the years.

Festuca varia, F. supina, Nardus stricta and other grasses form fragments of alpine meadows. Below is a list of some species appearing on the southern slope of 12–15° at 3,100 m (Mamisoni Pass, Greater Central Caucasus).

1. *Festuca varia*
Carex tristis
Poa alpina
Potentilla crantzii
Anthemis sosnovskyana
Campanula biebersteiniana
Anemone speciosa
2. *Festuca varia*
Alchemilla sericea

(continued)

Betonica macrantha
Poa alpina
Carex tristis
Anthemis sosnovskyana
Polygonum carneum
Taraxacum stevenii
Leontodon hispidus
3. *Festuca varia*
Poa alpina
Anthemis sosnovskyana
Aster alpinum
Veronica gentianoides
Campanula collina
4. *Festuca varia*
F. supina
Alchemilla caucasica
Carex tristis
Minuartia oreina
Veronica gentianoides
Polygonum carneum
Leontodon hispidus
5. *Festuca supina*
Minuartia oreina
Poa alpina
Anthemis sosnovskyana
Campanula saxifraga
Taraxacum stevenii

One can notice that the species of alpine meadows dominate but there appear plants from alpine carpets – *Taraxacum stevenii*, *Campanula biebersteiniana*, from scree habitats – *Anthemis sosnovskyana* and typical subnival landscapes – *Alchemilla sericea.* As in nanocoenoses fragments are formed by a small number of species, however here they occur in a large amount rather than in nanocoenoses. In more extreme environmental conditions are established fragments dominated by *Kobresia schoenoides* and *Carex tristis* appearing mainly on windy slopes of subnival zone and by alpine species – *Campanula biebersteiniana, Gnaphalium supinum, Taraxacum stevenii, Poa alpina* etc.

Fragments of alpine carpets or "Dikotylen-Teppiche" by Ellenberg (1996), Onipchenko and Blinnikov (1994) are established on moist places especially where snowbeds remain for the long time. On snow free places plants occur in better temperature conditions. There dominate espalier (*Minuartia aizoides, Sibbaldia semiglabra*) and rosette life forms (*Taraxacum stevenii*, *Plantago saxatilis* etc.)

Below is a list of species collected at 3,250 m on the southern slope of Mamisoni Pass:

Poa alpina
Carex medwedewii
Minuartia aizoides
Sibbaldia semiglabra
Taraxacum stevenii
Luzula pseudosudetica
Campanula biebersteiniana
Carum caucasicum
Saxifraga flagellaris
Mosses
Dicranum elongatum
Pohlia elongata
Pogonatum nanum
Tortulla tortuosa
Lichenes
Thamnolia vermicularis
Cetraria islandica
C. nivalis
Parmelia vagans etc.

Vegetation of alpine carpets often develops between large stones and screes, where stones keep snowbeds and on snow free places as well. Stones protect plants from strong winds and after heating give them supplementary warmth (Aleksandrova 1983; Nakhutsrishvili 1998a, b; Nakhutsrishvili et al. 1990). Below is a list of species occurring among stones in the subnival zone of mt. Elbrus at 3,350 m a.s.l.:

Minuartia aizoides	*Carum caucasicum*
Campanula biebersteiniana	*Saxifraga flagellaris*
Potentilla gelida	*Draba siliquosa*
Veronica minuta	*Alchemilla chlorosericea*
Festuca supina	*Senecio karjaginii*

Fragments of alpine carpets and turfy meadows as well are distinguished from real alpine meadows and carpets by appearing of species of subnival zone such as: *Veronica minuta, Saxifraga flagellaris, Senecio karjaginii* etc.

At the 3,100–3,350 m one can notice such extrazonal phenomenon as appearing of *Salix kasbekensis* on morains, slopes in Truso gorge (Kazbegi) and in Mamisoni Pass. We consider that it is a result of local foehns, when temperature increases while relative moisture decreases.

8.3.4 Life Forms

In the process of a long evolution, subnival plants have developed, along with metabolic and biochemical adaptive mechanisms, pronounced morphological qualities aiding in maximum warmth consumption, water loss prevention, protection against wind, drying up and withering caused by high temperatures of substratum surface. There are many plants with widely spread vegetative shoots (*Symphyoloma graveolens, Pseudovesicaria digitata, Scrophularia minima*) in this belt. There are also many plants with shoots creeping along the ground (*Veronica telephiifolia var. glareosa, Lamium tomentosum, Tripleurospermum subnivale*). Trailing and widely spread shoots of plant should be undoubtedly regarded as a form of adaptation to several environmental conditions, in particular, as a response to overheating of the substratum and to strong winds.

Dense-cushion plants (*Saxifraga subverticillata, S. ruprechtiana, S. moschata, Draba bryoides*) and loose-cushion plants (*Minuartia brotherana, M. inamoena*) are typical for the high-mountains of the Caucasus (Nakhutsrishvili and Gamtsemlidze 1984). Plant propagation ability is of special importance in the subnival belt with a thinned plant cover. This undoubtedly depends on the individual mobility of one or another species. 28 out of 91 investigated species were qualified as vegetative immobile (*Symphyoloma graveolens, Scrophularia minima, Jurinella subacaulis*) (Fig. 8.19). The other 63 species are vegetative mobile (*Minuartia inamoena, Veronica minuta, Cerastium polymorphum*). Few species belong to the type of vegetative highly mobile species, such as the plants with slight horizontal mobility. Perhaps this regularity can be accounted by severity of these ecotopes and the corresponding high degree of insufficiency in vegetative mobility of species.

Two groups of plants can be singled out in accordance with the nature of leaf downiness. The first group includes plants with undowny leaves (45 species): *Phryne huetii, Corydalis emanueli, Viola minuta.* In the second group are plants with downy leaves (46 species): *Alchemilla sericea, Aetheopappus caucasicus, Lamium tomentosum, Cerastium kasbek.* Downiness of plants protects them from strong solar radiation and from sever overheating of leaves. Among the plants with undowny leaves there are species with succulent or semisucculent types of leaves. For example, *Eunomia rotundifolia, Pseudovesicaria digitata, Tripleurospermum subnivale* and all species of *Saxifraga*, as well as the species with twisted leaves (*Festuca supina, Trisetum buschianum, Arenaria lychnidea*).

Eighty three species of investigated subnival plants of the Central Caucasus appeared herbaceous. The other eight species (*Sibbaldia parviflora, Ziziphora puschkinii*) are semi-shrubs (Fig. 8.19).

There are 81 species of polycarpics among the investigated plants, the other 10 species are monocarpics. The latter can be subdivided into perennial and biennial plants (Fig. 8.19). According to the data presented by Gamtsemlidze (1977), monocarpic annual plants are unavailable in the subnival belt, although some monocarpic plants of this belt (*Senecio sosnovskyi*) behave as annual monocarpic plants.

According to Raunkiaer classification, hemicryptophytes prevail in the subnival belt (Fig. 8.19).

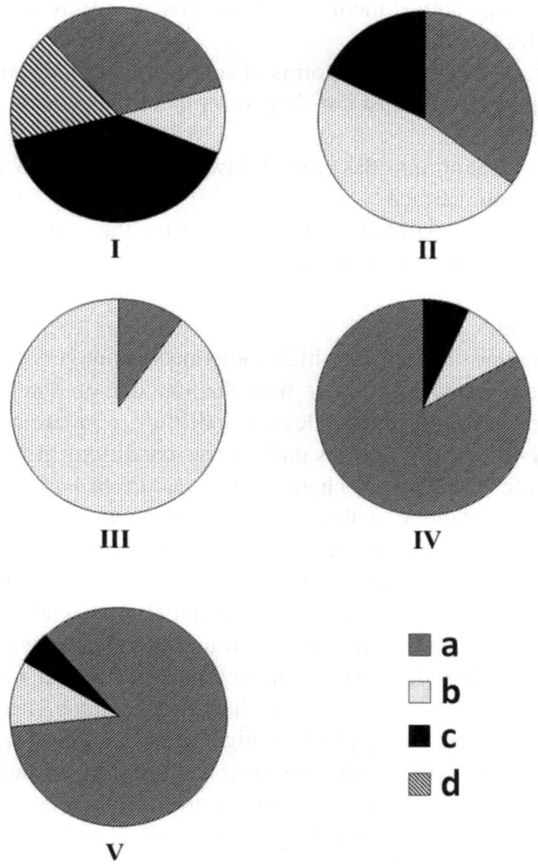

Fig. 8.19 Subnival plants comparison. Comparison of subnival plants with respect to the ecological-biomorphological indices: **I** - with respect to the way of specimen mobility: a - vegetatively immobile type; b - vegetatively poorly mobile type; c - vegetatively moderately mobile type; d - vegetatively intensively mobile type. **II** - with respect to the character of leaf downiness; a - plant leaves with downiness on both sides; b - plants with undowny leaves; c - plants with smooth upper and downy lower surface. **III** - with respect to shoot lignification: a - dwarf semi-shrubs; b - herbaceous plants. **IV** - with respect to lifespan: a - polycarpic; b - perennial monocarpic; c - biennial monocarpic. **V** - with respect to the type of plant reproductive bud position: a - hemicryptophyte; b - chamaephyte; c - geophyte (Nakhutsrishvili, Gamtsemlidze 1984)

Rhizome, taproot and rhizome-taproot plants are the most prevalent biomorphes in the subnival belt. There are much less mat-forming plants here. Tuberiferous plants are rarely found. Almost all the plants of this belt are characterized by intensive branching of underground organs near the surface of substratum.

Thus the subnival and nival belts are specified quite a large variety of life forms, notwithstanding their poor floristic composition; almost every species displays an independent life form. The variety of the life forms of plants, scattered on rocks, mounds and glaciers, enables them to use completely different ecological niches within a very restricted territory.

The most typical features of life forms of subnival belt plants are the following:

1. Out-spread, prostrate and cushion-shaped forms of the growth;
2. Miniaturization;
3. Not deeply penetrating into the ground downiness organs and widely branched near the substratum surface.

The out-spread and prostrated growth forms of the these plants promote:

(a) Maximum warmth consumption from the substratum surface warmed-up in sunny weather.
(b) Wind protection.
(c) Hibernating under snow cover which sometimes is not very thick.

Cushion-shaped plants form their microhabitat due to intensive orthotropic branching and spherical position of leaves and, thus, regulating the temperature relations and physiological processes under sever conditions of the belt.

Miniaturized life form "a very characteristic feature of humid alpine plants" is one of the important adaptive qualities for hibernating under a snow cover.

Downiness and succulence of leaves favour plant adaptation to the subnival and nival ecotopes with instable water relations protecting the stomatal apparatus from stress influence of wind and dry air, and establish an equalized water vapour concentration between leaves and the air around them (Larcher 1975; Cernusca 1976; Körner 2003). Besides leaf succulence ensures diurnal Crassulacean acid metabolism (CAM-plants), which under sharp temperature changes and in dry periods results in CO_2 uptake by plants at night.

Underground organs widely branched near a substratum surface provide active water absorbtion in the morning as soon as the soil begins to thaw. Another characteristic feature of underground organs of subnival and nival is their shallow rooting (about 15–20 cm) which protects them from cold lower layers (opposite data were obtained in the Alps (Körner 2003)). Even the longest underground organs at a certain level of depth (20–30 cm) begin to grow in horizontal direction (Nakhutsrishvili and Gamtsemlidze 1984).

8.3.5 Reproductive Buds

Buds of herbaceous plants may be bare (without special bud scales), protected (with spacial bud scales), and proleptic (with bud scales which in autumn begin to burst and new shoots leave their shells).

According to Gamtsemlidze (1979), in the subnival belt in severe winter conditions reproductive buds are covered by the residues of dead parts of plants and sheathes of dead leaves *(Jurinella subacaulis, Symphyoloma graveolens)*. But sometimes they occur in the surface layer of soil (*Lamium tomentosum, Cerastium polymorphum*) or deep in soil *(Delphinium caucasicum)*. Perennating buds of *Saxifraga sibirica* are covered by rosette-forming nodules which are partly in soil.

Reproductive buds of such wintergreen plants like *Saxifraga exarata, S. ruprechtiana* are protected by frame green leaves. A bud covering of plants which have protected proleptic buds is formed by solid bud scales and dead old leaves *(Scrophularia minima, Senecio karjaginii).*

It is characteristic that many subnival plants have stipules fitted closely to the base of stem and protect reproductive buds from damage *(Alchemilla sericea).*

Along with these adaptations for safe wintering of subnival plants snow cover protection is very important for reproductive buds. According to the data the biggest group of investigated plants is with bare buds (40 %), then the group with proleptic buds (35 %) and the smallest group is with protected buds (25 %).

The point is, what determines such a great amount of plants with bare reproductive buds in the subnival belt. There are a number of opinions on this problem. According to Serebriakov (1961): "The absence of bud scales in arctic and alpine plants is connected with fact that their scales are remetamorphosed into green assimilative leaves, which usually hibernate under the protection of a snow cover and partly under the protection of a cover of moss. One of the factors for the remetamorphosis of bud scales into green leaves in Tundra zone and high-mountain alpine belts was a considerable shortening of a vegetation period. Plants get an opportunity to extend the period of photosynthetic activity through remetomorphosis". Savile (1972) explains unprotection of reproductive buds with scales by the fact that hard scales hinder rapid spring development and hence, the plants are unable to adapt to short arctic vegetation period.

By the state of the next year shoot in reproductive buds at the end of the vegetation period, subnival plants could be divided into two groups: (1) plants with completely formed shoots in buds including not only vegetative organs but also inflorescenses and single flowers; (2) plants in the buds of which only the vegetative part of a shoot is formed, the inflorescence and flower germs appearing during the next vegetation period.

As much as 68 species out of 78 investigated subnival plants (88 %) possess generative shoots in reproductive buds which are born in the year before flowering (group 1) and only 10 species in the year of flowering (group 2).

According to the degree of flower development in reproductive buds the first group can be divided into three subgroups:

1. Next year flowers of reproductive buds are completely formed: sepales, a corolla, stamens, a pistil are developed (Fig. 8.20) (*Primula bayernii, Scrophularia minima*);
2. Next year flowers of productive buds are formed incompletely: sepales, petales, stamens are developed completely, the pistil is unformed (Fig. 8.21) (*Saxifraga sibirica, Potentilla gelida, Alchemilla sericea, A. chlorosericea, Draba siliquosa, Jurinella subacaulis*);
3. Inflorescenses and flowers of reproductive buds are not differentiated or start differentiating (Fig. 8.22) (*Senecio karjaginii, S. taraxacifolium, Tripleurospermum subnivale, Alopecurus glacialis, Delphinium caucasicum, Erigeron uniflorus*).

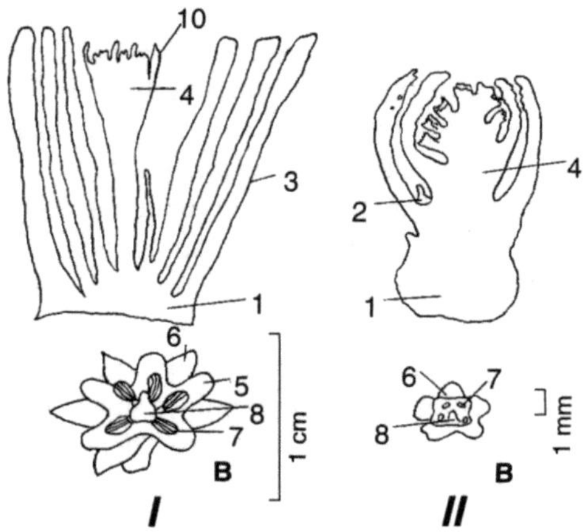

Fig. 8.20 *Primula bayernii*

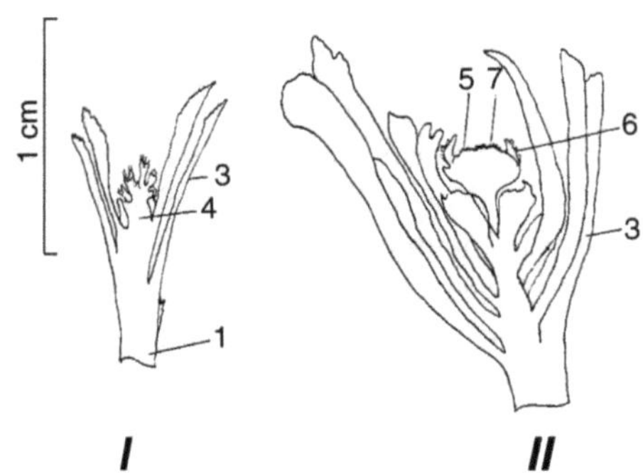

Fig. 8.21 *Alchemilla sericea* and *Potentilla gelida*

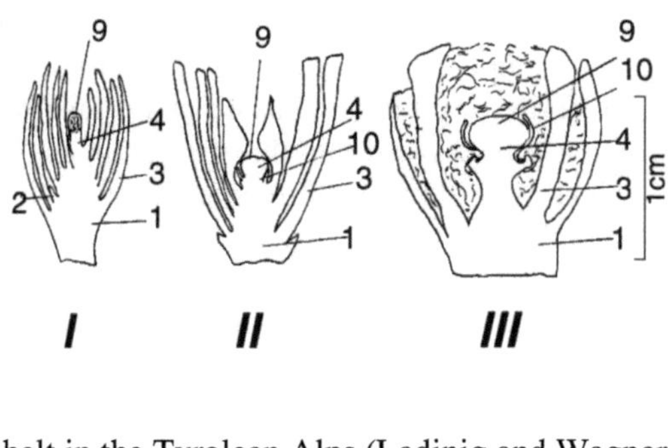

Fig. 8.22 *Alopecurus glacialis, Senecio taraxacifolius, S. karjaginii*

Profound studies of the subnival belt in the Tyrolean Alps (Ladinig and Wagner 2009; Steinacher and Wagner 2010) showed that flower differentiation in buds still continues except *Cerastium uniflorum*, in which flower emergence begins only 4–6 month prior to flowering and their opening is different in various species. For instance this period lasts 11 months in *Saxifraga oppositifolia*, 12–18 months in *Primula glutinosa* and *Saxifraga bryoides* and 24–26 months in *Ranunculus glacialis*. *R. glacialis* one of the flowering plants of the Alps extends to the highest zones. During the first year of the vegetation period stem buds appear at the base of old generative organs of the species, the flower cover and the first circle of stumen turn up before autumn. Next year the flower buds complete their development and in mid-August they turn into large closely situated flower buds (3–4 mm) and are ready to open with incompletely differentiated 1 year flower buds. Next spring the old buds open and the young ones continue to develop into new flower buds. *R. glacialis*, owing to such a periodical development of flowers retains the ability to flower every year in unfavourable conditions of vegetation.

Further development of flower buds of the plants, which are under snow cover during the whole summer stops for a while or at least slows down. Next year after the snow melts, these plants are able to bloom again, but their flowers will be plain and wretched. The plants (*R. glacialis*) covered by snow probably the germinated flowers perish. Such behaviour which is called "opportunistic" is contrary to the type of rapidly differentiating species (*Primula minima, P. glutinosa, Saxifraga oppositifolia*). Flower primordia of these species develop in July. In August, at the end of vegetation, the development of the originated flowers is completed, first of all among the plants located near the edge of glaciers. *S. oppositifolia* flowers immediately after the snow melting and just before the growth of vegetative parts begins. Under the conditions of habitat with much snow and extremely short period in comparison with the vegetative development is favourable, as due to high cold-resistance of flowers (10–15 °C), these species are able to enter a productive phase early and without any damage. Their seeds become ripe in cool summer. In the Pamirs (Steshenko 1967) it was also found that the majority of species have next year flowers and inflorescenses, originated in reproductive buds. High percentage of plants with previously originated inflorescenses and flowers can be observed in tundra (Nakhutsrishvili and Gamtsemlidze 1984).

8.3.6 Phenorhythmotypes

Phenorhythmotypes of the plants can be distinguished on the base of their phenological adaptation to environmental peculiarities. Taking into consideration a period favourable for vegetation in typical habitats, they can be distinguished by the behaviour of vegetative plants, that is, by the beginning and process of germination, by formation of leaves, their ageing, yellowing and defoliation, by the growth, germination and hibernation of buds (Borisova 1972; Larcher and Nakhutsrishvili 1982).

Below we give the phenorhythmotype classification of subnival plants.

A. Plants with Continuous Leafing

1. Evergreen plants: *Saxifraga ruprechtiana, S. scleropoda, S. exarata*
2. Summer-wintergreen plants:

Obligate summer-wintergreen plants: *Draba bryoides, Veronica minuta, Sedum tenellum, Saxifraga flagellaris, Antennaria caucasica*

Facultative summer-wintergreen plants: *Potentilla gelida, Viola minuta, Senecio karjaginii, Alopecurus glacialis, A. dasyanthus, Alchemilla sericea, Sibbaldia semiglabra, Ziziphora puschkinii, Tripleurospermum subnivale, Colpodium versicolor, Cerastium kasbek.*

B. Plants with Periodical Leafing

1. Summergreen plants with winter dormancy: *Scrophularia minima, Jurinella subacaulis, Phryne huetii, Chaerophillum humile, Valeriana saxicola, Silene marcowiczii, Minuartia inamoena, Anthemis sosnovskyana, Erigeron uniflorus, Delphinium caucasicum, Lamium tomentosum, Cerastium polymorphum, Symphyoloma graveolens, Aetheopappus caucasicus, Nepeta supina.*
2. Spring and early summergreen plants with the late summer-winter dormancy: *Corydalis alpestris, C. emanueli, Draba supranivalis, Primula bayernii, Arabis kazbegi.*
3. Late summergreen and autumn plants with summer-winter dormancy: *Crocus scharojanii, C. suworowianus.*

Among these groups summergreen plants with winter dormancy have the most widespreding (46.9 %), then come facultative summer-wintergreen (23.4 %), and spring-early summer-green with late summer-winter dormancy (9.4 %), the obligate summer-wintergreen (9.4 %), evergreen plants (7.8 %) and late summergreen-autumn (3.1 %) with summer-winter dormancy.

8.3.7 Temperature of Plants

Under extreme conditions of high-mountains the temperature of plants is one of the most important indicators of their life activity since it presents the true background for physiological processes when the amount of effective air temperature is small (Gorishina 1979; Körner and Larcher 1988).

Mountain plants show high leaf temperature and they should be included into the group of supratemperature plants which probably have high protoplasmic resistance. Such plants are identified as high temperature species (according to Kreeb 1974).

In the subnival belt of the Caucasus a very distinct surface temperature differentiation of species in accordance with surface temperature of plant organs is observed. Such species as *Potentilla gelida, Veronica minuta, Alopecurus dasyanthus* (Fig. 8.23), *Jurinella subacaulis*, have especially high temperature (29–30 °C). Leaf position of these plants in relation to the substratum is planophylic. All these species are spread to the upper line of flower plants. In comparison with other species *Potentilla gelida* is specified by the highest temperature in the West Pamirs, at the altitude of 4,750 m above sea level, as well as in the Caucasus (unpublished data by Gamtsemlidze and Chkhikvadze). Difference of leaf- air temperatures around the investigated plants amounts to 12–16 °C and 21–25 °C.

Fig. 8.23 *Alopecurus glacialis* (Photo O. Abdaladze)

Leaves of *Saxifraga moschata, S. kolenatiana*, semisucculent *Tripleurospermum subnivale* and leaves of plants with intensive downiness – *Anthemis sosnovskyana, Aetheopappus caucasicus* – warm-up relatively less. The lowest leaf temperature (23–25 °C) is recorded in coshion-plants – *Sibbaldia semiglabra, Minuartia inamoena.*

Plants with a high temperature of leaves also have the high temperature of flowers. In the investigated plants except cushion ones the temperature of flowers at noon is always lower than that of leaves. Very low temperatures of leaf surface (−4–6 °C) were registered early in the morning at great heights, when the plants turned into icicles. However, these plants thaw very slowly as they were shadowed by huge-boulders (otherwise they would have died). Temperature of leaves reached 28 °C and by 18–19 °C exceeded that of air. Subnival plants with prostrated and outstretched leaves over the substratum surface have to protect themselves against overheating when the temperature of leaves is much lower than that of substratum promoting normal water regulation and photosynthesis.

8.3.8 Structure of the Photosynthetic Apparatus

Structural adaptations related to photosynthetic activity play essential role in the adaptation of plants to habitat conditions, providing great ecological plasticity of the photosynthetic apparatus (Gorishina et al. 1985).

Among the investigated plants there are species with rather thin dorsivental lamellate leaves *(Oxyria elatior, Saxifraga sibirica)*, as well as with the isopalissade ones (*Astragalus kazbegi)* with the centric leaves of similar ones *(Arenaria lychnidea, Tripleurospermum subnivale)*, sometimes without clear differentiation of mesophyll between palissade and spongy tissues (*Ziziphora puschkinii*). The leaf structure of some species has distinct features of succulence. Leaf structure type is not closely connected with habitat conditions and shape of mesophyll cells. For instance, *O. elatior, S. sibirica* have narrow and elongated cells of palissade tissue, *S. juniperifolia*, has shorter and more rounded ones. As a rule, the cells of spongy tissue are round, the composition of the tissue being dense without big intercellural spaces. The cells of palissade tissue are packed very tight.

The investigated species are characterized by well-developed plastid apparatus. The mesophyll cells are filled densely with numerous chloroplasts, with their total volume being 10–30 % of the cell volume. The value of surface density of chloroplasts in leaves reaches 100,000–120,000/cm^2 accordingly the total surface chloroplasts is large. These peculiarities of the plastid apparatus are typical of the plants growing in habitats exposed to bright light. Such heliophilous plants grow in arid deserts, sea side littorals, deciduous forests in early spring as well as some in subalpine grasslands of the Caucasus. The values of the same indices in shade-requiring mesophytes are by an order more (Gorishina et al. 1985). Chloroplasts are usually small (*O. elatior* excluded) with very low chlorophyll content. This fact enables us to assume the presence of an ultrastructure with poorly developed system typical for the plants growing in the habitats exposed to bright-saturation. Chlorophyll content per unit of area and raw mass of the leaf is also not high (1.5–4 mg/dm^2 and 0.5–1.5 mg/g accordingly).

Light-requireness – "xerophilous" peculiarities of plants which can be called the "oreophytes" are expressed in different ways. They are best manifested in subnival oroxerophytes which inhabit rocks and talus and growing under the influence of substratum dryness and overheating. However, microhabitat conditions can be quite different even in these places. For example, the indices of plastid apparatus of *Saxifraga juniperifolia*, preferring more oversdaded and damp crevices, as a whole are mesophilous (small volume of chloroplasts in the cell and leaf, high concentration of chlorophyll in the chloroplast). The change reactions, e.g. under extreme conditions one of the orophil properties of the inner structure of leaves in plants, growing at the edges of a glacier or in hollows are expressed to a lesser degree. Only *Veronica telephiifolia var. glareosa*, having perennial green leaves and showing the trophorhythmotype of obligate-summer-winter green stands by itself (Larcher and Nakhutsrishvili 1982).

It was stated that the photosynthetical apparatus has considerable plasticity, manifested in the variability of leaf size and the indices of inner structure-surface density of chloroplasts etc. (Fig. 8.24). When the surface of a leaf varies in 3.5–4 times, the total reserve of plastids in the leaf varies in a more or less degree (1.5–2 times), as such diminution of the size of a leaf partially compensated by a considerable increase of the number of chloroplasts per unit of the leaf surface. That is to say, the plants growing in extreme conditions lose the reserve of chloroplasts in their leaves in a less degree in comparison with the loss of an assimilative surface that is important for the preservation of photosynthetic abilities of leaves. Chlorophyll content in chloroplasts differs substantially depending on a microhabitat (Fig. 8.25): both its total mass and concentration are slightly higher in species, growing in canopy.

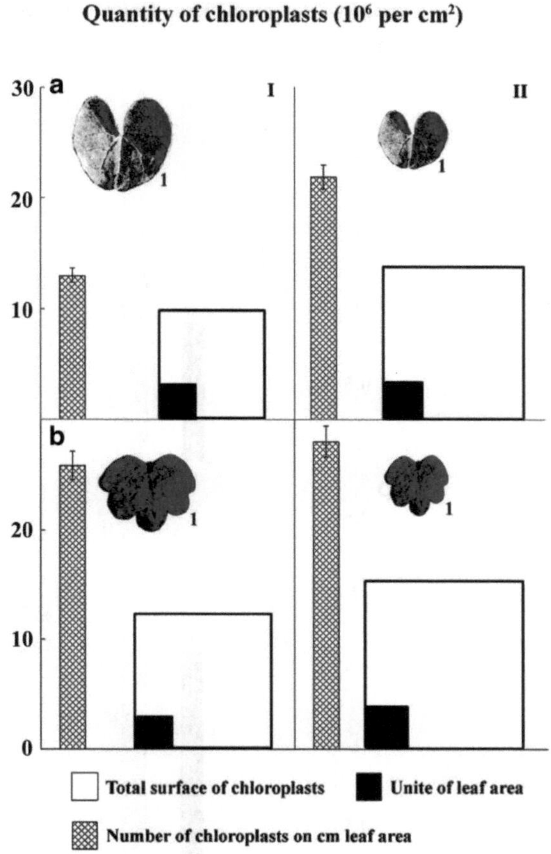

Fig. 8.24 Change of leaf characteristics

8.3.9 CO_2-Gas Exchange

In extreme conditions of high-mountains in the Caucasus the assimilative period of anthophytes is short (60–70 days). However, even during this time the assimilation period can be interrupted by frost or draught. Plant species respond to such extreme ecological conditions by different CO_2 gas exchange reactions, e.g. under the extreme conditions one of the groups of species had adaptable type of gas exchange, while in others the process of CO_2-gas exchange was rather stable (Nakhutsrishvili and Gamtsemlidze 1984; Abdaladze and Kikvidze 1991; Abdaladze 1994).

The plants of the first type are characterized by "impulsive" diurnal dynamics of net-photosynthesis (Fig. 8.26), frequent depressions of the CO_2 assimilation process, being accompanied by raising of stomatal diffusion resistance. The values of net-photosynthesis, diurnal productivity of photosynthesis, diurnal balance of CO_2-gas exchange and photosynthesis efficiency (high percentage of absorbed light energy stored in chemical links carbohydrates) should be considered as the

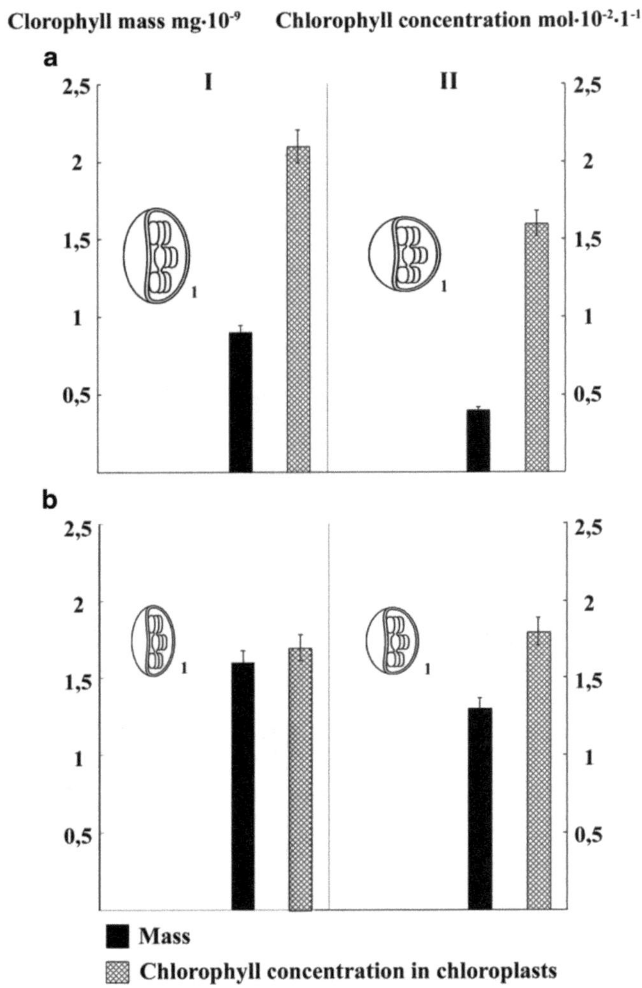

Fig. 8.25 Plant chloroplasts in different ecological microniches

distinguishing characteristic of the given type of CO_2 gas exchange. The plants showing the stable type of CO_2 gas exchange (succulent *Saxifraga kolenatiana)*, cushion-shaped (*Minuartia inamoena*) are characterized by absolutely opposite indices (Table 8.17).

Subnival plants, as well as other mountain plants are characterized by a very high net-photosynthesis. It should be noted that plants which are not distinguished by high intensity of photosynthesis (e.g. *Saxifraga kolenatiana)* demonstrate their economy by way of moderate use of assimilants that is displayed in very low intensity of dark respiration (Fig. 8.27).

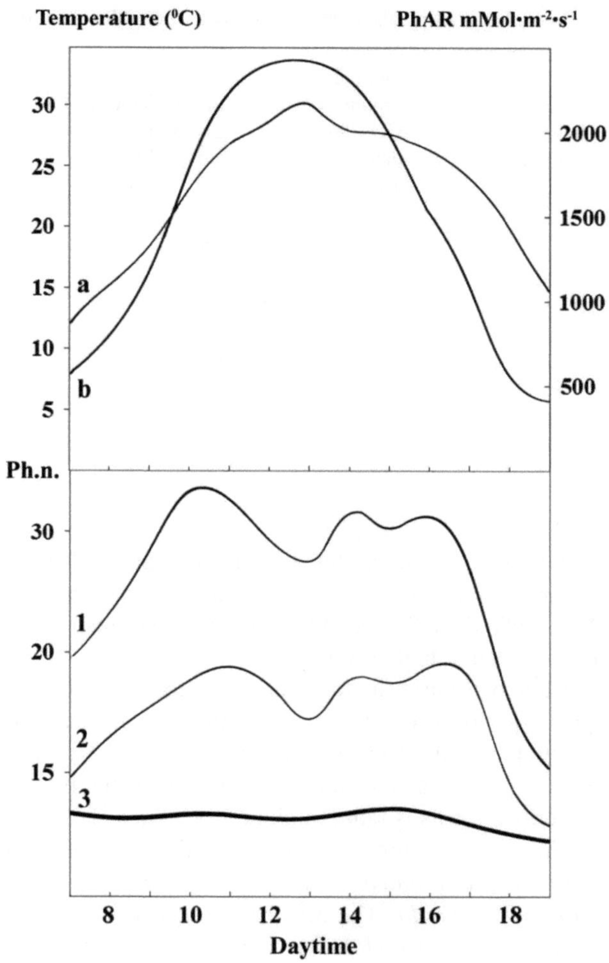

Fig. 8.26 Air temperature. PhAR and Net-Photosynthesis

Subnival plants are also characterized by wide temperature optimum of net-photosynthesis and well developed repair ability of assimilative apparatus (Moser 1973; Nakhutsrishvili and Gamtsemlidze 1984; Abdaladze 1994).

8.3.10 Diffusion Resistance of Leaves

"Xeromorphic" structure as well as trailing and cushion forms of plant growing promote stomatal apparatus regulation in the extreme conditions of high-mountains (Larcher 1975; Cernusca 1976; Körner 2003).

Table 8.17 CO_2-gas exchange of plants in the subnival belt of the Mamisoni pass at the altitude of 3,000 m s. m. (from Abdaladze 1994)

Species	Maximum net photosynthesis mg CO_2 dm^{-2} h^{-1}	Diurnal productivity mg CO_2 dm^{-2} per day	Daily balance of CO_2-gas exchange mg CO_2 md^{-2} 24 h	Photosynthesis efficiency %
Saxifraga kolenatiana	6,3	50	38	4,2
Arenaria lychnidea	42	277	230	20
Aetheopappus caucasicus	20	125	86	11
Alchemilla sericea	23	136	102	10
Minuartia inamoena	6,5	55	–	–

As is generally known, the higher the altitude the higher diffusive conductivity and duration of stomatal openings gaping. The main reason of this is lower pressure (Körner and Mayr 1980; Sanadiradze 1986; Körner 2003), responsible for the high level of plant evaporation at high altitudes.

Stomata diffusion resistance in the water relations and CO_2 gas exchange regulation of plants perform a very important function in the conditions of the subnival belt, where plants are isolated from each other and experience the impact of heavy dryness and heating of substratum. Diffusion conductivity in the majority of investigated subnival plants is rather high (1.5–1.7 cm s^{-1}).

As shown in Fig. 8.22, the highest level of conductivity of leaves with sharp fluctuations is observed in *Veronica minuta, Tripleurospermum subnivale, Senecio karjaginii* and *Potentilla gelida.* The *Saxifraga* species is characterized by low conductivity of leaves. No wonder that the values observed in this species *(Alopecurus dasyanthus)* are so small because of the ability of grasses to reduce the conductivity of water.

Majority of subnival plants are characterized by sharp diurnal water conductivity fluctuations. For example very often in the afternoon when high temperatures of substratum surface (50 °C and over), high levels of solar radiation (PhAR may extend up to 2,400–2,450 micromol m^2 s^{-1}) and in this respect the increased leaf temperatures (up to 300 C) occur, heavy depressions of water conductivity are observed. These depressions in different plants are manifested in different degrees.

Usually water conductivity in *Tripleurospermum subnivale* and *Veronica minuta* at noon compared to morning hours falls by 88–96 % (Fig. 8.28). Low midday deviations can be noticed in *Saxifraga* species, especially in *S. sibirica,* which at very high altitudes grows only under boulders. Almost flat curve of diurnal water conductivity was displayed by cushion-plants, e.g. *Sibbaldia semiglabra,* a dwarf

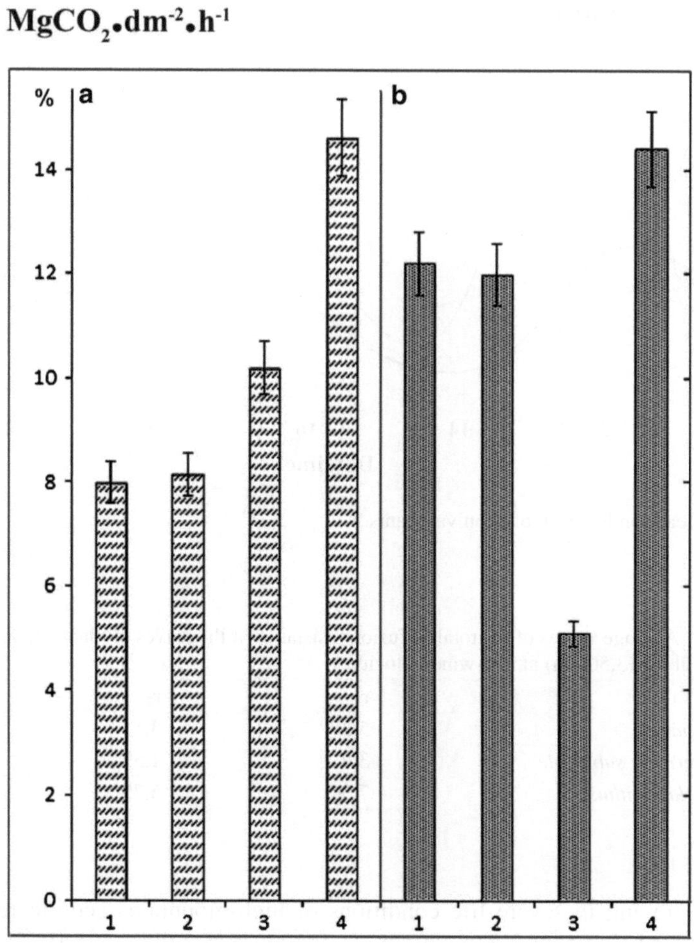

Fig. 8.27 Economical coefficient of Photosynthesis and dark respiration

subshrub. However at high altitudes in unclosed communities average diurnal conductivity of leaves falls.

Average values of total diffusion resistance of leaves in the subnival belt of the Elbrus (3,500 m a.s.l.) at low wind velocity.

Table 8.18 shows data on total water diffusion resistance (rt). So called leaf boundary layer resistance is very important for transpiration. Its share in investigated species except *Tripleurospermum subnivale* as compared to subalpine grasses is smaller than the share of stomatal resistance (rl). Thus in extreme life conditions (especially in the plants growing in open places exposed to strong winds) stomatal play the most vital role in water – relations of plants.

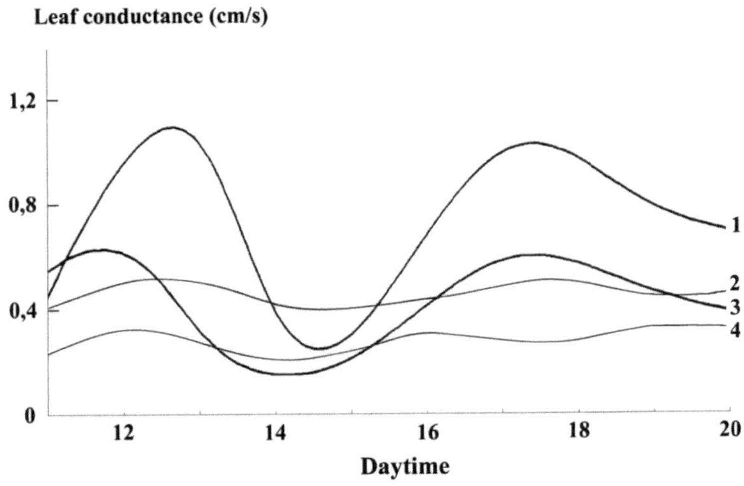

Fig. 8.28 Leaf conductivity of subnival plants

Table 8.18 Average values of the total diffusion resistance of the leaves of plants in the subnival belt of the Elbrus (3,500 m) at low wind velocity

Species	r_1	r_2	r
Veronica minuta	5.38	3.11	2.26
Tripleurospermum subnivale	3.6	1.52	1.50
Alopecurus dasyanthus	7.30	5.75	1.54

Plants growing in severe life conditions of high-mountains acquire reactivity potential after being exposed to extreme temperature (Semikhatova 1965; Larcher 1980; Zalensky 1982). This has been confirmed by the results obtained in the course of observations of diffusion resistance of stomata in morning hours, i.e. after the impact of negative temperature of air (−2.50 C), of substratum surface (−1.6 °C) and of leaf (−3.0 °C). In all species after sunrise (630–1,010 μmol m^{-2} s^{-1}) and with the increase of air temperature (10 °C) and substratum temperature (5.0 °C) leaf conductivity amounts up to 0.75–1.15 sm s^{-1}. Trailing plants, such as *Veronica minuta*, *Tripleurospermum subnivale* and the grass *Alopecurus dasyanthus* actively respond to the solar radiation and rise of temperature. Whereas *Senecio karjaginii* (plant with erect stalk and equally distributed leaves) reacts less actively and its leaf conductivity stays below 0.21 sm s^{-1}.

8.3.11 Leaf Water Potential of Plants

Plants of subnival belt possess high transpiration and water potential values (Nakhutsrishvili and Gamtsemlidze 1984). The question is whether water balance becomes equalized in the extreme conditions of high mountains. Larcher (1983) considers that water potential reflects any alterations in water balance as its tension immediately provokes loss of turgor and subsequent growth of water potential.

Very often the increase of PhAR and temperature of the boundary air layer and leaves results in the reduction of water potential of most subnival species. *Saxifraga* species are characterized by rather stable water potential, though.

A certain correlation has been observed between conductivity and water potential of leaves. Minimum conductivity of leaves corresponds to the minimum values of their water potential. In the diurnal course of water potential, we have noticed an interesting regularity, which is well manifested in the plants of highlands, after the morning temperature stress (Fig. 8.29). Water potential undergoes gradual reduction until quite low values at noon. It restores in the second half of the day, due to the growth of diffusion resistance.

It is interesting to compare leaf water potential of plants in the Central Caucasus and in the Pamirs. Observations carried out by Gamtsemlidze and Chkhikvadze (unpublished data) in the region of the Fortambek Glacier (north-west region of the Pamirs) at the height of 4,000–4,750 m above sea level, show that water potential of leaves in the Pamirs is 2–3 times lower than the same index in plants of the Caucasus (the Elbrus). The minimum water potential of plants in the Pamirs is 2.8–3.3 MPa. Water potential in *Potentilla gelida* in the Caucasus amounts to 5–11 MPa, whereas in the Pamirs it is 1.1–2.5 MPa. It should be noted, that at high altitudes of the Pamirs (4,750 m) water potential of leaves, which is rather low, can still reduce to 3.0 MPa, e.g. in *Waldheimia glabra*.

8.3.12 Discussion and Conclusions

Subnival and nival ecosystems (viz. microecosystems) scattered among rocks and mounds, consist of a small number of anthophyte species. Thanks to adaptation they can survive in severe life conditions. High-mountain inhabitants obtain the quality to resist such factors as low temperature and frosts during the vegetation period, warmed-up substratum, hurricane wind and besides, low carbon dioxide content and the pressure of water vapour in the air. These plants have a short vegetation period and a great deal of chionophiles stay under a thick snow cover during 18 months.

Apparently, evolution of plants here, in contrast to the alpine and subalpine belts is aimed at the selection of species with high resistance against heavy physical conditions; here competition seems to be less important for the selection of species.

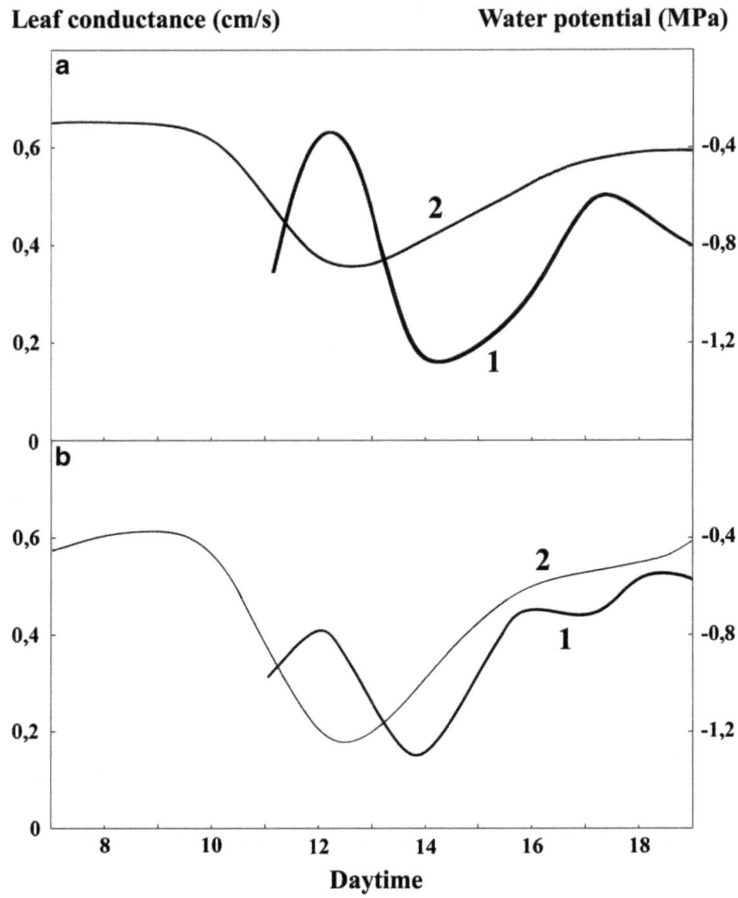

Fig. 8.29 Relations between leaf conductance and water potential

Severe climatic and edaphic conditions have a strong impact on the structure of plant cover. Vertical structure of plant groupings in the subnival and nival belts is characterized by the absence of stratum differentiation and unclosed horizontal structure. Unlike other altitudinal belts root systems here are closed only in case of penetration by one plant into the tufts of another, which leads to the formation of micro-and ultramicrogroupings.

Subnival belt displays a great variety of life forms, despite a restricted number of species; in extreme life conditions almost every species demonstrates an individual life form with its typical ecophysiological reaction to environmental changes. Probably, this accounts for the fact that various species of plants individually adapt to their existence within the same ecotope.

Analysis of life forms of plants of this belt revealed the prevalence of widely prostrate and cushion-plants with downy and semisucculent leaves. The subnival belt is also characterized by vegetative-movable plants, which form clumps and patches. They provide the security of widely represented rock-talus ecotopes. According to the Raunkiaer system the most frequent life forms here are illustrated by hemicryptophytes and chamaephytes. These life forms were caused by the adaptation of plants to the subnival ecotopes, which are subject to the impact of day-night temperature fluctuations and strong wind, the latter is of particular importance.

High-mountain biomes are very rich in diverse life forms, especially the mountains of the torrid zone. Every mountain system and every altitudinal belt involves specific life forms which are typical only for them. For example, mountains of the torrid zone are characterized by rosette-like megaphytes, large bunches of tussock grasses, downy and thorn plants; in arid high mountain regions prevail cushion and prostrate dwarf subshrubs and shrubby plants. Humid high mountain regions of temperate zones represent trailing shrubs and dense turf plants, small delicate grass species, and cushion-shaped, rosette-like, trailing and prostrate herbaceous plants which dominate in the extreme conditions of the subnival and nival belts. In alpine tundra the prevalent life forms are shrubs and small shrubs. In general, humid high-mountain regions of temperature and high latitude zones are characterized by tiny life forms as compared to the tropical and arid (low latitude) mountains (Nakhutsrishvili 1981)

Plants growing at high altitudes of both tropical and temperate zones are characterized by "xeromorphism" particularly by strong downiness, succulence, twisting of leaf, its thickening and substitution by thorn, etc. (Nakhutsrishvili 1974; Larcher 1975). These properties were regarded as a reaction of plants to water deficiency or poor mineral nutrition. The recent data advised by Larcher (1975) suggest that small pressure of water vapour and the state of stomata along with other factors, determine the "xeromorphism" of plants. Water deficiency, caused by freezing of soil is observed in different high-mountain biomes (Alps, Andes, Caucasus, etc.) occurring only at night or in the morning. During the rest of time, plants absorb water from soil.

Körner (2003) observed that the degree of stomatal opening is subject to the effect of dry air of high-mountain regions; various species react to dry winds (foehns) differently. Any increasing of the resistance of leaf boundary layer through their downiness, succulence, twisting protects the stomatal apparatus from strong influence of dry air and wind and establish a equalized concentration of water vapour of leaf and its environmental water. As a result of the scientific observation conducted by Cernusca (1976) it was found that the analogous effect was observed in the plants which have cushion and trailing growth forms. According to Larcher (1975) the ecological benefit of so called xeromorphic structure of high-mountain plants is in the levelling of air humidity fluctuations and, thus in maintaining (through the system of stomatal apparatus) the diffusion resistance of carbon dioxide gas and water at a comparatively optimum level.

Generative tissue, reproductive buds in particular, has decisive significance for plant adaptation to severe environmental conditions. Subnival plants are mainly characterized by bare and proleptic opening buds. Only a small number of species have closed ones. The majority of subnival plants have perennating reproductive buds in which flowers and inflorescences appear beforehand, that is one of the essential signs of adaptation of these plants to a very short period of vegetation.

The analysis of plants distribution in their habitat with different thickness of snow cover showed (Nakhutsrishvili and Gamtsemlidze 1984) that the number of species considerably increased from the minimum snow zone to the maximum one. The least number of species accounted for the maximum snow zone. Unlike other altitude belts in the subnival belt, there are the least number of chionophobous plants. In this belt as well as in tundra so called greenhouse effect is apparent in spring. It must be mentioned that in such "greenhouse" the vegetation period begins about 2 weeks earlier in comparison with one in the conditions of an open biotope.

The beginning of plant vegetation depends mainly on the periods of snow cover disappearance and so fluctuations have especially pronounced character at the beginning of it. Mass vegetation of plants in the subnival belt is observed only from the end of June till the end of September and the beginning of October (depends on snow fall). The first flowering species appear in the only in the second 10-day period of June reaching their culmination in the second half of July. The flowering period in the given belt is observed foe a rather long time, that lasts till the second half of September. During this period plants can hardly be found in the alpine and often in the lower subalpine belt.

Even the smallest territories of the subnival belt of the Greater Caucasus were rich in various phenorhythmotypes. Both, evergreen with summer-wintergreen species can be found there. Summergreen plants with winter dormancy were the prevailing phenorhythmotype of the subnival belt. Then comes facultative summer-wintergreen plants.

We came to the conclusion that plants growing at the highest altitudes were presented by high-temperature species which could survive in extreme conditions. Leaf temperature in sunny weather rises on average, than the ambient temperature and falls by the same degrees below the temperature of substratum surface over which leaves and shoots of subnival plants prostrate. The transpiration favours the hydrothermal regulation of plants not only in steppes and deserts but also in subnival conditions. In differentiation of plants in respect of temperature of their organs, the architectonics of plants, particularly the pattern of the position of their leaves in relation to the sun rays and highly warmed-up substratum is of high decisive importance. On the basis of example it is possible to consider the plants of cushion-shaped life from which is widely distributed almost in every biome with extreme conditions. According to our investigation the plants of such life form as opposed to many species are not overwarmed and at noon in sunny weather their leaf temperature is always optimal. Probably, such homeostasis is due to a ultramicrohabitat created by a specific form of growth of these plants. Their temperature conditions become favourable for active water exchange and carbon assimilation. The same function can be carried out by turf. Some of the plants not typical for the subnival belt and without such adaptive mechanisms, turn yellow and wither form intensive solar radiation in August.

Tolerance limits as to the temperature stress in the most subnival and nival plants are rather high, e.g. these plants may be due to their recover ability, leave the stress very quickly in spite of the fact that during a day leaf temperature varies from −2, −5 °C to +30, +34 °C. Specifically, diffusion resistant of leaves decreases, water exchange and assimilation activity of plants are restored.

Temperature conditions of plants and their environment have a large extent an impact of diffusion resistance of leaves. Almost all the species investigated were found to be sensitive to stomata regulation. Plant with erect stocks are subject to winter influence to a greater degree and during a day period their stomatal conductivity varies. As it is apparent, basically in extreme conditions of subnival belt, diffusion resistance of leaves has to perform the function of water balance and CO_2 gas exchange of plants.

As a whole a water relations of subnival plants is well regulated although the cases of considerable rise of water low level, considerable reduction of water content in leaves and water potential, elevation of osmotic pressure and water conductivity reduction are observed in many species. That is why all the subnival plants which have been investigated, except succulents, are included in the group with hydrolabile species and at the same time, with euryhydrous (Nakhutsrishvili and Gamtsemlidze 1984).

It is quite natural that peculiarities of water relations has an influence on carbon exchange of plants. For instance, the plants with low water content and high degree of water shortage are characterized by low intensity of photosynthesis. Midday depression of photosynthesis often observed at noon are mainly the result of increased diffusion resistance of leaves caused trees environmental conditions. Night frosts and midday depressions are the factors which limit day and seasonal course of photosynthesis. However, due to the recover ability of plants, their assimilation activity is restored rapidly after temperature stresses.

One of the most important strategies of plant survival in unfavourable conditions of high mountains is a high temperature optimum of net-photosynthesis. On the basis of our work we concluded that a short period of assimilation in extreme high-mountain conditions may be sufficient for a considerable carbon accumulation.

Thus, there are species with absolutely different mechanisms of adaptation to the habitat conditions within small territories, they are micro and ultramicrogroupings.

During a long period of evolution, high mountain plants have acquired ability to survive and reproduct in extreme conditions. The principal mechanism of plant adaptation are the following:

- Growing species via the medium of both caintagius and individual distribution of plants among rock, latus and glaciers.

- Formation of various specialized forms of growing resulted in the protection against the impact of unfavourable habitat conditions.
- Completion of seasonal development cycle of plants in a short period of vegetation.
- Early origination of almost all plant organs including all parts of flowers in reproductive buds.
- Final formation of leaves after more favourable ecological conditions coming.
- High leaf resistance to sharp daily fluctuations of temperature.
- High temperature of leaves with their thermoregulation capacity, which assure active assimilation activity and water exchange of plants.
- Regulation of physiological processes which manifest itself due to diffusion resistance of leaves.
- High temperature optimum for photosynthesis.
- High level of photosynthesis in the period between unfavourable weather conditions and low dark respiration in plants with poor photosynthetic activity resulting in economical expenditure of assimilants.
- Well developed recover capacity in plants after stress impact.
- The tendency to leaf succulence formation for optimum keeping of water balance and assimilation activity.
- Long under-snow-cover stay of plants due to considerable accumulation of organic matter reserve in underground organs.
- The variety of life forms, pheno-and trophorhythmotypes within the strongly limited territory demonstrating the species ability to occupy and develop their ecological niche.

In the conclusion we find it necessary to describe some distinctive characteristics of plants growing in extreme conditions of high-mountain regions of the Caucasus.

There are small or no such plants as bushes, dwarf shrubs, representative of sedge family in the subnival and nival ecosystems as opposed to alpine meadows having common border line with them. In such conditions root systems are closed very seldom and territory of unclosed vegetation cover; polygonal nature of relief –that is the most characteristic property of the alpine belt-is hardly marked. As the altitude increases the activity of species decreases. In the subnival and especially nival belts the species activity are not enough or inactive (e.g. *Cerastium kazbek*).

It must be outlined that the analogous regularity has been revealed by Aleksandrova (1983) as related to polar desert. This geobotanical region as well as the subnival and nival belts is characterized by small number of flowering plants, the absence of bushes and representatives of sedge family, the uncloseness of vegetation cover of species activity and etc.

When passing from the alpine belt to the subnival sharp changes of plant growth form is observed (cushion plants and prostrate ones predominate). In extreme conditions of high mountain regions the species with "xeromorphic" features hold a dominating position as opposed to mesophilous plants which are typical for the alpine belt.

Subnival and nival belt plants unlike the species of alpine belt are characterized by peculiar rhythmological features and physiological processes. For instance, they differ from each other by impulsive and peaked curves of efflorescence and two pronounced maxima of flowering. Diversity of aspects of ultramicrogroupings and aggregations is observed in the subnival belt even in the late summer phase of plant development while in the subalpine and alpine belts only some flower species. (Nakhutsrishvili and Gamtsemlidze 1984). Aspects change in vegetative groupings is expressed neither in extreme conditions of polar deserts nor in the nival belt (Aleksandrova 1983). Interrelations of physiological processes with environmental factors is more tense in subnival plants than e.g. in alpine or subalpine plants in the conditions of closed grass cover. In the first case, in spite of high species ability to adapt to the unfavourable conditions, strong overheating of assimilative surface of plants, photosynthesis depression and sharp increasing of leaf diffusion resistance are often observed.

Unlike the alpine belt, plant ecological stress in the subnival and nival belts becomes heavier due to the absence of closed vegetation cover (basically grass sods) promoting the development of various homeostatic mechanisms.

The species growing in the subnival and nival belts in the period of their formation experienced the climatic stresses caused both by sharp fall of temperature and its rise. Undoubtedly, these plants were witnesses of the period of volcanic activities when they had to adapt themselves to the existence conditions of extreme with high concentration of carbon dioxide gas, mobile and highly warmed-up substratum, etc.

Only such high adaptation ability of these plants to comparatively big number of external factors helps them to survive in conditions of extreme press. The plants adapted to such wide range of changing ecological conditions are the representatives of the most ancient flora of high mountain regions of the Caucasus.

8.4 High Mountain Plant Community Diversity

Vegetation of the humid climate orobiomes is characterized by high sensitivity and its irrational use frequently results in severe erosion. One example in point is Kazbegi region where plant diversity and endemism level are fairly high (Sakhokia and Khutsishvili 1975). Over 20 % of the Caucasian flora and all vertical zones of the Caucasus are represented here (Fig. 8.30); particularly high plant diversity is observed in the upper montane and subalpine zones.

The upper montane zone of Kazbegi region experiences strong anthropogenic stress, in particular, it is negatively influenced by urban areas, overgrazing, conversion of traditional hay fields to pastures, etc. Plants of almost all the vertical zones can be found there. A range of contrasting phytocoenoses is diverse including steppe, rock and scree tragacanth vegetation, of the one hand, and wetlands and meadows, on the other (Table 8.19). The majority of the coenoses are found on plain areas but some of them also occur on quite steep slopes. Meadows Bromopsieta variegatae, Hordeta violaceum Festuceta variae are distinguished by high specific diversity. Moist meadows Blysmo-Phragmitetum and eroded meadow-steppe Bromopsio-Festucetum contain relatively small number of species (Table 8.19).

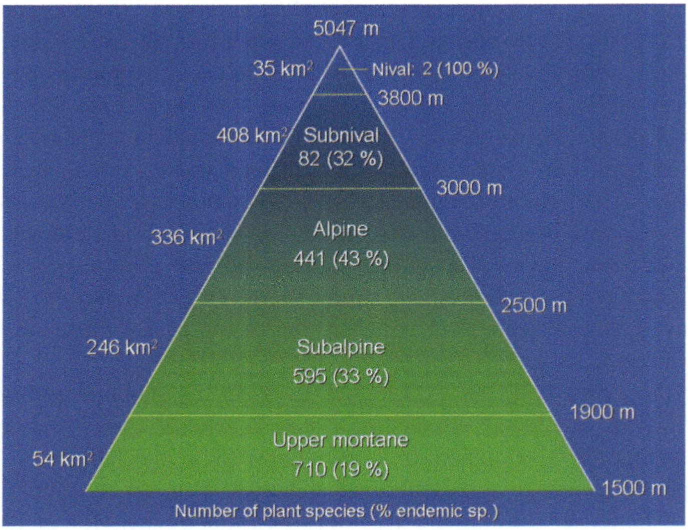

Fig. 8.30 Plant species diversity by altitudinal gradient in Kazbegi mountains

In subalpine zone Festuceta variae, Agrostio-Bromopsietum, Calamagrostieta are the coenoses with high number of species. It should be noted that meadow-steppes Festuceta ovinae, Koelerieta are also characterized by high floristic diversity (Table 8.19). The subalpine tall herb vegetation Heracletum is very poor in species.

In alpine zone the general vegetation background is similar to that in subalpine zone; floristically rich as well as poor coenoses can be found there. For instance, still Festuceta variae is the coenosis distinguished by high specific diversity. Nardeta is another floristically rich coenosis. Festuceta supinae, Kobresieta capiliformis contain small number of species.

Rhododendretum caucasici also does not contain high number of species.

Statistic analysis has shown that correlation between species richness (at 25m^2) and elevation is weak negative ($r=0.3$, p,0.05, $n=39$). It has revealed as well weak relationships between species richness and the following habitat parametres: community height, plant cover, slope inclination ($r=0.3$, $r=0.4$, and $r=0.4$, respectively; $p<0.05$, $n=39$). The number of species (per area unit) is abruptly dropped only above 2,800–2,900 m a.s.l.

8.5 A Comparative Analysis of the High Mountain Vegetation of the Caucasus and the Alps

The characteristic feature of high mountain vegetation of the Caucasus is the development of xero – and hemi-xerophytic vegetation, i.e. feather-grass steppes, steppe meadows, thorny cushion plant communities and meadows of the Lamiaceae family at the treeline and in the lower reaches of the alpine zone. These communities are not found in the high mountains of the Carpathians, the Alps

Table 8.19 Diversity of plant community of central Caucasus according to the altitude gradient

Plant communities	m.a.s.l. (m)	Inclination (°)	Exposition	Community height (cm)	Cover %	Number of species per 25 m
Upper montane belt						
Bromopsio – *Festucetum valesiacae*	1,740	0	O	40	70	25
Blysmo – *Phragmitetum australis*	1,740	0	O	60	80	20
Agrostio tenuis – Bromopsietum variagatae	1,750	0	O	40	95	20
Deschampsio cespitosae – Hordeetum violaceae	1,750	0	O	50	95	21
Triseteto rigidum – Festucetum ovinae	1,750	0	O	70	60	23
Festuco ovinae – Bromopsietum variegatae	1,750	5	W	50	90	21
Astragaletum denudatus	1,750	40	OS	70	60	23
Deschampsio flexuosae – Equisetetum arvense	1,750	0	O	30	90	21
Agrostio tenuis – Festucetum pratensis	1,800	0	O	100	100	30
Hordeetum violaceae	1,800	5	N	60	100	22
Hordeetum violaceae	1,800	2	S	100	100	28
Trisetetum rigidae	1,800	5	NO	50	70	27
Bromopsietum variegatae	1,800	15	NO	50	100	43
Poo pratensis – Hordeetum violaceae	1,800	2	W	70	100	27
Trifolio fontanum – Hordeetum violaceae	1,800	2	SW	70	100	27
Agrostio tenuis – Bromopsietum variegatae	1,850	25	NW	40	100	34
Agrostio tenuis – Bromopsietum variegatae	1,850	5	W	120	100	25
Hordeetum violaceae	1,850	5	W	160	100	25
Hordeetum violaceae	1,850	2	W	120	100	22
Hordeetum violaceae	1,850	3	W	15	70	22
Festucetum valesiacae	1,850	25	NW	60	95	38
Hordeetum violaceae	1,870	3	S	80	100	41
Agrostio tenuis – Bromopsietum variegatae	1,870	2	S	40	80	38

(continued)

Table 8.19 (continued)

Plant communities	m.a.s.l. (m)	Inclination (°)	Exposition	Community height (cm)	Cover %	Number of species per 25 m
Festuco ovinae – Agrostietum tenuis	1,870	0	O	50	80	23
Festucetum ovinae	1,897	15	S	70	95	33
Subalpine belt						
Cariceto meinshausenianae – Festucetum variae	1,900	30	W	70	100	32
Festucetum ovinae	1,900	5	S	50	100	27
Festucetum ovinae	1,900	5	S	20	90	26
Agriostio tenuis – Bromopsietum variegatae	1,900	5	S	50	100	29
Agrostio tenuis – Bromopsietum variegatae	1,900	5	S	50	95	33
Bromopsio variegatae – Agrostietum tenuis	1,900	25	SWS	50	90	33
Bromopsio variegatae – Koelerietum luerssenii	1,900	20	SWS	15	90	26
Bromopsio variegatae – Festucetum ovinae	1,900	30	S	50	70	23
Agrostio tenuis – Bromopsietum variegatae	1,900	20	SW	50	90	26
Agrostio tenuis – Bromopsietum variegatae	1,900	0	O	40	90	28
Agrostio tenuis – Bromopsietum variegatae	1,900	1	O	50	95	30
Bromopsio variegatae – Agrostietum tenuis	1,900	5	S	50	100	29
Agrostio tenuis – Bromopsietum variegatae	1,900	0	O	50	100	54
Bromopsio variegatae – Festucetum ovinae	1,900	10	S	40	70	21
Bromopsio variegatae – Festucetum ovinae	1,900	3	SO	40	80	23
Agrostio tenuis – Bromopsietum variegatae	1,900	15	NW	50	80	22
Bromopsio variegatae – Scabiosetum caucasici	1,900	5	S	50	100	22
Bromopsio variegatae – Agrostietum tenuis	1,910	3	S	40	90	41

(continued)

Table 8.19 (continued)

Plant communities	m.a.s.l. (m)	Inclination (°)	Exposition	Community height (cm)	Cover %	Number of species per 25 m
Agrostio tenuis– Festucetum ovinae	1,940	10	NW	50	90	43
Agrostio tenuis – Bromopsietum variegatae	1,945	3	SW	50	90	52
Cariceto meinshausenianae – Festucetum variae	1,950	35	NW	80	100	28
Cariceto meinshausenianae – Festucetum variae	1,950	40	SW	75	80	26
Cariceto meinshausenianae – Festucetum variae	1,950	30	WN	80	100	39
Cariceto meinshausenianae – Festucetum variae	1,950	35	SW	70	80	47
Cariceto meinshausenianae – Festucetum variae	1,950	30	WN	80	40	29
Nardetum strictae	1,950	9	NO	40	100	27
Nardetum strictae	1,950	10	NO	40	100	29
Agrostio tenuis – Bromopsietum variegatae	1,950	15	S	40	100	33
Helictotrichono asiaticus-Bromopsietum variegatae	1,950	10	S	40	95	28
Helictotrichono asiaticus-Bromopsietum variegatae	1,950	5	S	40	95	35
Bromopsio variegatae – Koelerietum luerssenii	1,950	15	S	50	95	25
Agrostio tenuis – Bromopsietum variegatae	1,950	20	SW	50	100	38
Agrostio tenuis – Bromopsietum variegatae	1,950	10	S	40	95	28
Agrostio tenuis – Bromopsietum variegatae	1,950	15	S	40	100	35
Cariceto meinshausenianae – Festucetum variae	1,950	35	NW	70	100	39

(continued)

Table 8.19 (continued)

Plant communities	m.a.s.l. (m)	Inclination (°)	Exposition	Community height (cm)	Cover %	Number of species per 25 m
Medicagini glutinosae – Bromopsietum variegatae	1,950	15	S	40	90	19
Bromopsio variegatae – Agrostietum tenuis	1,950	5	S	30	90	27
Festucetum ovinae	1,960	30	SW	40	90	40
Agrostio tenuis – Bromopsietum variegatae	1,970	30	N	70	100	38
Calamagrostietum arundinaceae	1,970	30	N	70	100	38
Festucetum ovinae	1,970	25	S	50	90	38
Agrostio tenuis-Bromopsietum variegatae	1,970	25	SW	50	90	38
Festucetum ovinae	1,980	20	S	50	95	27
Bromopsio variegatae – Festucetum ovinae	1,980	20	S	50	95	27
Bromopsio variegatae – Festucetum ovinae	1,980	20	S	50	90	24
Agrostio tenuis-Bromopsietum variegatae	1,980	20	S	30	95	27
Agrostio tenuis-Bromopsietum variegatae	1,980	20	S	50	95	24
Festucetum ovinae	1,987	20	SW	25	90	34
Festucetum ovinae	2,000	30	S	40	80	32
Agrostio tenuis-Bromopsietum variegatae	2,000	30	S	45	80	32
Nardetum strictae	2,000	15	SO	80	100	25
Bromopsio variegatae – Agrostietum tenuis	2,000	5	WNW	50	100	24
Festucetum ovinae	2,000	30	S	45	80	32
Agrostio tenuis-Bromopsietum variegatae	2,000	3	W	40	90	39
Pulsatillo – Festucetum ovinae	2,000	20	SO	40	90	30
Vicietum alpestris	2,000	5	S	40	90	20
Cariceto meinshausenianae – Festucetum variae	2,000	30	W	50	100	17

(continued)

Table 8.19 (continued)

Plant communities	m.a.s.l. (m)	Inclination (°)	Exposition	Community height (cm)	Cover %	Number of species per 25 m
Agrostio planifoliae – Bromopsietum variegatae	2,000	25	N	50	90	34
Bromopsio variegatae – Festucetum ovinae	2,000	15	S	40	90	20
Cariceto meinshausenianae – Festucetum variae	2,010	25	NO	60	100	28
Cariceto meinshausenianae – Festucetum variae	2,010	25	SW	60	100	39
Kobresio capilliformis – Festucetum ovinae	2,020	35	NO	20	90	42
Bromopsio variegatae – Kobresietum capilliformis	2,020	35	NO	30	90	42
Agrostio tenuis-Bromopsietum variegatae	2,025	20	SW	40	90	34
Festucetum variae	2,045	35	S	20–80	90	32
Bromopsio variegatae–Festucetum ovinae	2,050	10	SWS	50	100	26
Bromopsio variegatae–Festucetum ovinae	2,050	10	SWS	30	100	25
Bromopsio variegatae–Festucetum ovinae	2,050	10	SWS	10	90	28
Koelerietum luerssenii	2,050	5	SO	70	95	31
Koelerietum luerssenii	2,050	5	NW	40	95	26
Festucetum pratensis	2,050	0	O	80	100	28
Cariceto meinshausenianae – Festucetum variae	2,050	35	SW	100	90	21
Cariceto meinshausenianae – Festucetum variae	2,050	35	S	80	80	32
Cariceto meinshausenianae – Festucetum variae	2,050	30	NW	80	90	32
Cariceto meinshausenianae – Festucetum variae	2,050	30	NW	80	90	23
Festuco variae – Dryetum caucasici	2,050	35	NW	70	90	35
Festucetum ovinae	2,050	15	S	40	90	33
Festucetum ovinae	2,050	10	SO	30	95	20

(continued)

Table 8.19 (continued)

Plant communities	m.a.s.l. (m)	Inclination (°)	Exposition	Community height (cm)	Cover %	Number of species per 25 m
Pulsatilletum violaceae	2,050	35	SW	50	90	28
Festucetum ovinae	2,050	25	SW	25	90	37
Pulsatillo violaceae - Bromopsietum variegatae	2,050	35	SWS	50	90	28
Helictotrichono asiaticus–Bromopsietum variegatae	2,050	10	NWN	50	100	31
Bromopsio variegatae – Koelerietum luerssenii	2,050	5	S	70	95	31
Bromopsio variegatae – Koelerietum luerssenii	2,050	5	NW	40	95	26
Bromopsio variegatae – Koelerietum luerssenii	2,050	20	SO	30	95	24
Agrostio tenuis-Bromopsietum variegatae	2,050	15	S	30	90	27
Agrostio tenuis-Bromopsietum variegatae	2,050	0	O	60	100	27
Agrostio tenuis-Bromopsietum variegatae	2,050	3	N	40	100	24
Agrostio tenuis-Bromopsietum variegatae	2,050	10	NW	50	100	31
Agrostio tenuis-Bromopsietum variegatae	2,050	10	SO	30	95	31
Agrostio tenuis-Bromopsietum variegatae	2,050	5	SO	70	95	31
Agrostio tenuis-Bromopsietum variegatae	2,050	5	NW	40	90	26
Agrostio tenuis-Bromopsietum variegatae	2,050	20	SO	30	95	24
Nardo strictae – Deschampsietum flexuosae	2,050	5	W	30	90	24
Vicietum alpestris	2,050	15	SW	30	100	17
Caricetum meinshausenianae	2,050	20	W	25	100	27

(continued)

Table 8.19 (continued)

Plant communities	m.a.s.l. (m)	Inclination (°)	Exposition	Community height (cm)	Cover %	Number of species per 25 m
Cariceto buschiorum – Bromopsietum variegatae	2,050	0	O	30	90	20
Cariceto buschiorum – Bromopsietum variegatae	2,050	3	S	40	90	21
Cariceto meinshausenianae – Festucetum variae	2,055	30	S	100	80	16
Cariceto buschiorum – Festucetum ovinae	2,060	0	O	20	60	20
Helictotrichono asiaticus – Bromopsietum variegatae	2,060	15	NW	20	100	28
Festuco ovinae – Koelerietum luerssenii	2,060	0	O	20	90	20
Agrostio tenuis-Bromopsietum variegatae	2,060	15	NW	20	90	28
Trifolio ambiguum – Agrostietum planifoliae	2,080	30	NO	60	90	47
Trifolio ambiguum – Agrostietum planifoliae	2,080	30	NO	30	90	47
Nardetum strictae	2,100	20	S	50	100	31
Nardetum strictae	2,100	20	S	35	95	31
Trifolio ambiguum – Bromopsietum variegatae	2,100	25	SW	40	90	33
Agrostio tenuis – Poetum alpinae	2,100	30	N	60	90	28
Caricetum meinshausenianae	2,100	25	NW	30	90	33
Cariceto tristis – Festucetum variae	2,100	40	NW	60	80	39
Poetum alpinae	2,100	0	O	40	80	25
Festuco supinae – Sibbaldietum semiglabrae	2,105	0	O	20	70	40
Cariceto meinshausenianae – Festucetum variae	2,110	25	NO	70	100	50
Cariceto meinshausenianae – Festucetum variae	2,110	25	NO	60	100	50

(continued)

Table 8.19 (continued)

Plant communities	m.a.s.l. (m)	Inclination (°)	Exposition	Community height (cm)	Cover %	Number of species per 25 m
Cariceto meinshausenianae – Festucetum variae	2,120	35	NW	60	100	43
Agrostio tenuis – Festucetum ovinae	2,140	12	N	20	100	37
Festucetum supinae	2,150	20	SO	15–30	100	31
Kobresietum capilliformis	2,150	0	O	30	95	25
Kobresietum capilliformis	2,150	5	N	20	90	29
Blysmo – Deschampsietum flexuosae	2,150	0	O	40	80	25
Bromopsio variegatae – Kobresietum capilliformis	2,150	5	SO	30	100	27
Helictotrichono asiaticus – Festucetum ovinae	2,175	10	NO	30	90	44
Bromopsio variegatae – Kobresietum capilliformis	2,175	5	N	15	100	26
Anemonetum fasciculatae	2,190	20	SSW	30–40	95	22
Nardetum strictae	2,200	20	S	20	90	40
Nardetum strictae	2,200	20	S	30	90	40
Heracleetum sosnowskyi	2,200	0	O	200	80	7
Vaccinio myrtilli – Rhododendretum caucasici	2,200	20	N	60	100	16
Cariceto tristis – Anthennarietum caucasici	2,200	20	N	20	80	20
Alchemillo elisabethae – Kobresietum persica	2,200	15	SW	40	50	29
Trifolio ambiguum – Bromopsietum variegatae	2,200	10	SW	40	90	55
Kobresietum capilliformis	2,200	25	SO	20	100	28
Kobresio capilliformis-Nardetum strictae	2,200	10	N	25	100	27
Kobresio capilliformis-Bromopsietum variegatae	2,200	15	NO	30	100	38

(continued)

Table 8.19 (continued)

Plant communities	m.a.s.l. (m)	Inclination (°)	Exposition	Community height (cm)	Cover %	Number of species per 25 m
Cariceto meinshausenianae – Festucetum variae	2,220	25	O	60	90	31
Agrostio planifoliae – Nardetum strictae	2,220	5	NO	50	100	23
Deschampsietum cespitosae	2,250	0	O	100	100	13
Nardetum strictae	2,250	15	W	100	100	27
Nardetum strictae	2,250	5	SWS	25	95	27
Vaccinio myrtilli – Nardetum strictae	2,250	25	NW	30	90	39
Chamaesciadio acaule– Kobresietum capilliformis	2,250	5	SO	20	100	32
Daphno glomeratae – Festucetum ovinae	2,300	20	N	30	100	41
Pulsatillo violaceae– Festucetum ovinae	2,300	10	SW	20	90	33
Cariceto tristis – Festucetum variae	2,300	30	NW	60	90	40
Kobresietum capilliformis	2,300	10	SW	15	100	25
Nardeto strictae– Vaccinietum myrtilli	2,300	15	NO	35	100	30
Cariceto tristis – Festucetum variae	2,300	35	SO	50	80	27
Deschampsio flexuosum – Nardetum strictae	2,300	20	NO	30	100	47
Cariceto tristis – Festucetum variae	2,300	35	S	50	80	27
Cariceto meinshausenianae – Festucetum variae	2,330	20	S	30	80	25
Cariceto meinshausenianae – Festucetum variae	2,350	30	S	80	70	17
Nardetum strictae	2,350	25	SW	70	100	31
Bromopsietum variegatae	2,350	15	SO	50	100	34
Alchemillo sericatae – Festucetum ovinae	2,350	5	SO	30	90	40
Cariceto medwedewii – Nardetum strictae	2,390	5	NW	40	100	28
Alpine belt						
Sibbaldietum semiglabrae	2,400	10	N	80	100	26

(continued)

Table 8.19 (continued)

Plant communities	m.a.s.l. (m)	Inclination (°)	Exposition	Community height (cm)	Cover %	Number of species per 25 m
Nardetum strictae	2,400	0	O	40	100	24
Festuco variae – Rhododendretum caucasici	2,400	25	N	60	100	42
Pulsatillo violaceae – Festucetum ovinae	2,400	15	SW	20	90	26
Festuco variae – Rhododendretum caucasici	2,400	30	NW	70	100	38
Kobresietum capillifromis	2,410	0	O	15	80	22
Nardetum strictae	2,420	15	W	20	95	20
Sibbaldio parviflorae – Caricetum tristis	2,450	5	NW	30	90	25
Cariceto tristis – Festucetum variae	2,450	20	SW	50	100	43
Vaccinio myrtilli – Rhododendretum caucasici	2,450	20	N	30	100	27
Kobresietum capilliformis	2,450	0	O	15	90	26
Taraxacetum stevenii	2,500	5	N	20	90	18
Bromopsio variegatae – Caricetum tristis	2,500	5	NO	40	100	28
Deschampsio flexuosae – Nardetum strictae	2,500	5	N	40	100	35
Sibbaldio parviflorae – Nardetum strictae	2,500	5	N	25	100	35
Festuco variae – Dryadetum caucasici	2,500	40	N	70	100	33
Festuco supinae – Nardetum strictae	2,500	10	SO	20	100	41
Cariceto tristis – Alchemilletum elisabethae	2,550	5	NW	20	100	26
Sibbaldio semiglabrae – Caricetum medwedewii	2,550	0	O	15	90	16
Kobresietum capilliformis	2,550	10	SO	12	90	34
Sibbaldietum semiglabrae	2,560	10	N	30	100	19
Pulsatillo violaceae-Festucetum ovine	2,560	25	SO	40	100	23
Cariceto meinshausenianae – Festucetum variae	2,560	25	SO	40	100	39

(continued)

Table 8.19 (continued)

Plant communities	m.a.s.l. (m)	Inclination (°)	Exposition	Community height (cm)	Cover %	Number of species per 25 m
Festuco variae – Dryadetum caucasici	2,560	25	SO	40	100	23
Cariceto tristis – Festucetum variae	2,600	30	NO	50	100	31
Deschampsio flexuosae – Nardetum strictae	2,600	5	N	40	100	24
Cariceto tristis – Festucetum variae	2,600	20	NW	60	90	32
Festuco supinae – Kobresietum persica	2,600	20	SO	20	90	24
Deschampsio flexuosae – Nardetum strictae	2,650	5	SO	30	90	29
Festuco supinae – Kobresietum persica	2,700	0	O	15	80	13
Rhododendretum caucasici	2,700	25	N	30	80	22
Kobresio capilliformis – Caricetum tristis	2,700	15	SO	15	90	40
Poo alpinae – Nardetum strictae	2,700	15	SO	20	100	37
Festuco supinae – Caricetum tristis	2,700	15	NO	20	90	16
Festuco supinae – Nardetum strictae	2,700	15	SO	20	100	37
Rhododendretum caucasici	2,750	40	N	30	70	14
Kobresio capilliformis – Festucetum supinae	2,800	10	N	20	80	24
Kobresietum schoenoides	2,840	0	O	30	80	15
Festucetum variae	2,850	35	S	40	90	23
Kobresio schoenoides – Caricetum tristis	2,850	30	S	30	70	13
Kobresio schoenoides – Festucetum variae	2,850	15	SW	20	90	28
Kobresio capilliformis – Festucetum supinae	2,850	5	SO	15	90	22
Cariceto tristis – Festucetum supinae	2,900	10	N	30	70	16
Kobresio capilliformis – Festucetum supinae	2,900	0	O	15	80	29
Kobresietum capilliformis	2,950	20	S	15	95	20
Festucetum supinae	2,980	20	S	20	80	25

(continued)

Table 8.19 (continued)

Plant communities	m.a.s.l. (m)	Inclination (°)	Exposition	Community height (cm)	Cover %	Number of species per 25 m
Subnival belt						
Subnival –Nanocoenoses	3,200	0	O	5	20	5
Cerastium kazbek-Alopecurus dasyantus	4,000	0	O	5	1	2

and the Pyrenees (Franz 1979; Grabherr and Mucina 1993; Ellenberg 1996; Grabherr 1997; Ozenda 1985, 2002; Nakhutsrishvili and Ozenda 1998; Burga et al. 2004; Erschbamer et al. 2010; Pauli et al. 2011; Korner 2012).

In the Alps the treeline is mainly formed by conifers (*Pinus cembra, P. mugo, Larix decidua, Picea abies*) (Ozenda 1985, 2002), while in the Caucasus the same zone is dominated by broad-leaved trees (*Betula litwinowii, B. medwediewii, Acer trautvetteri, Sorbus caucasigena, Quercus macranthera*).

There is no analogue of Colchic per-humid relict vegetation in the Alps.

The tall herbaceous vegetation has no analogue in the Alps. Although some species of tall-herb vegetation are common to the Alps and the Caucasus (e.g. *Aconitum nasutum, Doronicum macrophyllum, Telekia speciosa*), the tall-herb vegetation of the Caucasus is unique in the number of species, level of endemism and stature. The *Betula-Adenostyles* tall-herb community of Ellenberg (1996) has little in common with the tall-herb vegetation of the Caucasus, which ecologically is more of an equivalent of the *Alnus viridis* scrub of the Alps (Nakhutsrishvili and Ozenda 1998).

Meadows of *Calamagrostis arundinacea, Festuca varia, Hordeum violaceum,* and *Trisetum flavescens* occur both in the Caucasus and in the Alps; however, in contrast to the Alps, *Arrhenatherum elatius* is not a characteristic grassland species in the Caucasus.

Dry meadows with *Bromopsis variegata* and diverse herb meadows with *Anemonastrum fasciculatum, Betonica macrantha, Geranium ibericum, Inula orientalis, Ranunculus caucasicus, R. oreophilus, Scabiosa caucasica,* and *Trollius patulus* are widespread in the Caucasus but are absent from the Alps. *Kobresia (= Elyna) myosuroides* (Alps) and *Kobresia capilliformis* (Caucasus) rush haths, *Woronowia speciosa-Carex pontica* (Caucasus, on limestines) and *Carex firma* (Alps, on limestones) sedge heaths can be regarded as vicarious communities.

The most important difference between the high mountain vegetation of the Alps and the Caucasus is that contrasting nature of the Caucasus plant cover including the per-humid relict plants (Colchis), on the one hand, and high mountain steppe and xerophilous scrub, on the other.

Many subnival plants in the Caucasus are local endemic species, which have phylogenetic relationships with species from Asia Anterior–South-West Asia or Asia Minor (Nakhutsrishvili and Gagnidze 1999), but unrelated to the species of the (sub) nival zone of the Alps. The vascular plant species of the Alps are widely distributed and occur in most European high mountain areas, while local endemism prevails in the Caucasus (Nakhutsrishvili 2003).

Hydrophytic Vegetation

9

Bog vegetation is a typical component of phytolandscapes of Georgia. It covers an area of approximately 200 ha. Bogs are very common on Colchic Lowland and South-Georgian Volcanic Upland, populating chiefly lowlands and subalpine belt, whereas in mid-mountain and alpine zone they are rare. The commonest bogs in Georgia are eutrophic, of which mesotrophic and oligotrophic formation occupy limited area. Under the special conditions of subalpine belt of the Greater Caucasus oligotrophic pit bogs are developed on crusts of weathering of different rocks.

Vertical distribution of *Sphagnum bogs* in West Georgia comprises all altitudinal belts from sea level to the subalpine zone; *Sphagnum* bogs are very rare in alpine belt with its severe climatic conditions. These bogs are not to be found at the altitudes below 2,000 m in East Georgia due to the aridity of climate of this region.

Distribution area of *Sphagnum* bogs (meso-oligotrophic types) involves western parts of the Greater and Lesser Caucasus, where they occur in the zone of *Picea-Abies* forests and mesic beechwoods. The following obligatory helophytes are characteristic of West Georgia: *Scheuchzeria palustris, Carex flava, Viola palustris, Cardamine seidlitziana, Heracleum apiifolium, Sredinskya grandis, Rhynchospora alba, Drosera rotundifolia, D. intermedia* and *D. anglica*.

Bog vegetation of Georgia is similar to that of North Eurasia, though one cannot find here many characteristic components of northern peat bogs. *Sphagnum* bogs in Colchic Lowland and Ajara Mts. are characterized by the presence of such relic species as *Rhododendron ponticum, Rh. luteum, Vaccinium arctostaphylos, Frangula alnus, Osmunda regalis*, etc. This type of vegetation is rich in species with disjunctive and wide area of distribution. It is moderately manifested as endemics of Georgia and the Caucasus (8 %). Among the above-mentioned endemics the following should be mentioned: *Hibiscus ponticus, Heracleum apiifolium, Rhynchospora caucasica, Scirpus colchicus, Sredinskya grandis, Carex szovitsii, Cardamine seidlitziana, Primula luteola, P. lazica, Oenanthe abchazica, Pycreus colchicus* and *Swertia iberica*.

Hydric tall herbaceous vegetation occurs mainly on lowlands and Volcanic Upland of the South Georgia. Its upper limit here is 2,000 m. One of the most

G. Nakhutsrishvili, *The Vegetation of Georgia (South Caucasus)*, Geobotany Studies, DOI 10.1007/978-3-642-29915-5_9, © Springer-Verlag Berlin Heidelberg 2013

common formations of this group is Phragmiteta australis. Typheta is considered to belong to the above-mentioned group. It is dominated by *Typha latifolia, T. angustifolia* or *T. laxmanii*. Schoenoplecteta with its phytocoenoses dominated by *Schoenoplectus lacustris* or *S. tabernaemontani* attracts a special attention.

Cladieta marisci is populated on the lowlands and foothills of the Caucasus. The dominant *Cladium mariscus* is a thermophilous relic species; its upper limit lies at an altitude of 800 m.

Glycerieta arundinaceae and *Digraphieta arundinaceae* manifest rare formations of hydric tall herbaceous bog vegetation.

In Georgia the hydric low herbaceous bog vegetation extends from sea level up to 2,300 m. In general it occupies a negligible area. Equiseteta, which is dominated by *Equisetum heleocharis, E. palustris and E. ramosissimum* species, is confined to lowlands and low mountain zone.

Hippuris vulgaris is one of rare obligatory helophytes.

Sparganieta is one of the most common formations of hydric low herbaceous vegetation. The monodominant phytocoenoses of this formation are dominated by *Sparganium polyedrum* or *S. simplex*.

Butometa umbellati, according to the published data, occupies limited areas in Georgia. In mountains it is represented just by one association – *Butometum umbellati*.

Cariceta dichoandrae, Cariceta acutiformis, Cariceta elatae, Cariceta cespitosae, Cariceta wilnicae and Magnojunceta. Cariceta elatae is rare in Georgia, as well as Cariceta caespitosae, which is spread in South Georgian Volcanic Upland and Ajara Mts. Cariceta wilnicae is confined to Javakheti Upland; it extends from 2,000 to 2,100 m

Magnojunceta differ from above-mentioned *Carex* communities; it is dominated by *Juncus effusus* or *J. inflexus*.

Eutrophic bog vegetation including herbs with short rhizomes is very common for the subalpine belt of Georgia and the Caucasus; at the same time some formations are also characteristic of the alpine zone, for example Cariceta dacicae (Cariceta transcaucasicae).

Cariceta kotschyana are typical for alpine belt of the Greater Caucasus; they extend from 2,000 up to 2,900 m a.s.l.

Blysmeta compressi are spread in the mountains of Georgia; the dominant species here is *Blysmus compressus*. The species of *Sphagnum* are not present among the coenoses of this formation.

Eumesotrophic bog vegetation, including herbs with short rhizomes, is also characteristic of subalpine belt. It plays an important role in the process of accumulation. The typical representative of this formation is *Cariceta canescenti*. Moss synusium is dominated by *Calliergon stramineum* and *C. giganteum*, rarely by *C. richardsonii* and *C. cordifolium*.

Cariceta inflatae, which are confined to the subalpine belt are a typical syntaxon of eumesotrophic bog vegetation of Georgia.

Cariceta lasiocarpae are a typical formation of peat bog vegetation. *Carex lasiocarpa*, distributed in Holarctis, bears an important function in the process of peat accumulation. Cariceta lasiocarpae are very abundant in North Eurasia and North America.

Cariceta chordorrhizae are widespread in North Eurasia, while it is rarely found in the Caucasus. The leading species *Carex chordorrhiza* is an obligatory helophyte.

Cariceta songoricae are very rare in the Caucasus; their distribution area involves marl schists and limestones of Kavkasioni.

Menyantheta trifoliae occupy mountain bog surfaces; their upper limit is situated at the altitude of 2,250 m.

Molinieta litoralis are restricted to the Colchic Lowland and is an important component of peat bog phyto-landscape.

Mesotrophic herbaceous bog vegetation is mainly distributed in West Georgia from sea level up to the subalpine belt. It participates in the process of peat accumulation. Formations of this group include Scheuchzerieta palustrae, a rare component in the vegetation of the Caucasus. It occurs only in Svaneti (West Georgia), at 1,700–2,100 m.

Cariceta limosae are a typical representative of mesotrophic herbaceous bog vegetation. The dominant *Carex limosa* is an obligatory helophyte developed on peat bog surface.

Cariceta irriguae extend from 1,750 to 2,400 m in the mountains of Georgia.

Eriophoreta vaginati are a typical formation of mesotrophic herbaceous bog vegetation; they are widely distributed in the taiga zone of Eurasia and occupy a limited area in Georgia and the Caucasus.

Rhynchosporeta caucasici are another interesting formation of the above-mentioned vegetation; they are typical for Colchic peat bogs.

Mesotrophic *Sphagnum* bogs mostly occur in the mountains of West Georgia.

The area of distribution of *Sphagnum* bogs comprises lowland and mountainous parts of West Georgia. Peat accumulation here is limited provided for the unfavourable physico-geographical conditions. In western part of Kavkasioni *Sphagneta magellanici* are confined to the upper limit of *Picea/Abies* forests. A frequent component of meso-oligotrophic bog vegetation is Sphagnetum magellanici cariosum. Its moss synusium is characterized by a high presence of *Sphagnum angustifolium* and *S. amblyphyllum.*

Typologically, *Sphagneta palustrae* are one of the most interesting formations of oligotrophic *Sphagnum* bog vegetation. They play an important part in the vegetal landscapes of Colchic Lowland. Their moss synusium is dominated by *Sphagnum palustre*. Frequent associates of dominant species are *Sphagnum apiculatum, S. papillosum, S.contrale, S. amblyphyllum,* etc.

Sphagnetum palustrae caricoso-moliniosum occupies a vast area in Colchic Lowland. Herbaceous synusium is dominated by *Molinia litoralis* and *Carex lasiocarpa*, rarely by *Carex acutiformis*.

Sphagnetum palustrae scheuchzeriosum extends from 1,700 up to 2,100 m, while Sphagnetum palustrae rhynchosporosum albae occupies areas above 1,500 m. These communities are distributed in West Georgia.

Fig. 9.1 *Nymphaea alba*, the Lesser Caucasus (Tsalka). (Photo O. Abdaladze)

Sphagneta angustifolii are considered to belong to the group of formations of oligotrophic *Sphagnum* bogs. They extend from 1,600 up to 2,000 m and in some parts even to 2,500 m.

The distribution area of *Sphagneta imbricati* is situated in the southern part of Colchic Lowland. Sphagnetum imbricati moliniosum litoralis belongs to the commonest coenoses. The above-mentioned communities involve the following constant species: *Molinia litoralis, Carex lasiocarpa, Rhynchospora caucasica, Drosera rotundifolia, Osmunda regalis,* etc.

Sphagneta fusci, characteristic of Eurasia and North America, occur in the Caucasus and Georgia, in particular. The dominant *Sphagnum fusci* is a rare component of the Caucasian moss flora. Another example of rare communities is Sphagneta acutifoliae.

Bog vegetation with arboreal components occupy considerable areas in the lowlands of West Georgia, whereas they are poorly represented in the mountains. *Alnus glutinosa* plays an important part in the communities distributed in lowland; its frequent associate in mountains is *Alnus incana.*

The *Alnus glutinosa-A. incana-Frangula alnus-Phragmites australis* complex covers the surfaces of several bogs in the mountains of West Georgia.

Bog vegetation with shrubs is very rare in Georgia. Mainly it extends on the mountains between 1,700 and 2,000 m. These communities can be illustrated by Saliceta varioherbosa and Saliceta herboso-sphagnosa (Kimeridze, 1992).

Aquatic vegetation of Georgia is also worth of a brief account. This type of vegetation still needs to be researched though.

According to Kolakovsky (1961), *Lemna* groupings occur on stagnant waters, ditches and pools of Colchic Lowland. Typical plants of this grouping are *Lemna minor, Spirodela polyrhiza, Salvinia natans.*

Fig. 9.2 *Nuphar luteum*, Colchis. (Photo O. Abdaladze)

Hydrocharis communities group consists of plants, which occur on standing waters and among the thickets of tall aquatic herbs. Two ecological types can be outlined within this communities group. The first type is manifested by *Ruppia* community, which occurs in water lagoons along the coast and island salt lakes This community consists of *Ruppia spiralis*, *R. maritima*, *Zostera marina*, etc. The second ecological type is represented by freshwater plants, such as *Potamogeton lucens*, *P. natans*, *P. crispus*, *Myriophyllum spicatum*, *Nymphaea alba* (Fig. 9.1), *N. candida*, *Nuphar luteum* (Fig. 9.2) and *Trapa colchica*. The rare comminity of *Utricularia vulgaris* and *Aldrovanda vesiculosa* are to be found in the lakes Bebe-Syr and Anishkhara (Abkhazeti).

Middle mountain aquatic vegetation of West Georgia is floristically very poor. *Callitriche verna* is the only typical aquatic plant of the alpine zone of Georgia (Kimeridze 1992).

10 Synanthropic Vegetation

The synanthropic vegetation involves three large groups: (1) the segetal plant communities of cultivated habitats; (2) the ruderal vegetation formed by plants growing in the vicinity of dwellings, roadsides, disturbed lands, etc.; and (3) the pastoral communities group of pasture weeds.

10.1 Segetal Vegetation

The segetal plant communities of Georgia are extremely diverse floristically and ecologically and should be tentatively specified into the following five groups.

10.1.1 Grain Crops

The segetal communities of grain crops are very rich in species. Over 700 species grow in corn, wheat and barley fields. Examples of the most trouble some weeds are *Avena ludoviciana*, *Lolium rigidum*, *Secale segetale*, *Sinapis arvensis*, *Rumex crispus*, *Convolvulus arvensis*, *Setaria pumila* and *Agrostemma githago*.

Corn fields in Georgia are manifested by *Setaria viridis*, *Amaranthus retroflexus*, *Cynodon dactylon* and especially by *Sorghum halepense*.

In montane zones of the Caucasus and Georgia, in particular, floristic composition of segetal communities is enriched by the plants, which frequently occur on mountain slopes. Of these plants the following may be mentioned: *Bunias orientalis*, *Vaccaria hispanica*, *Melampyrum arvense* and *Bifora radians*.

10.1.2 Cultivated Subtropical Plants

In the West Georgia large areas are occupied by the tea plantations. It should be noted that these plantations provided 95 % of the tea produced in the former USSR.

G. Nakhutsrishvili, *The Vegetation of Georgia (South Caucasus)*, Geobotany Studies, DOI 10.1007/978-3-642-29915-5_10,

Paspalium digitaria (North American native), *Digitaria sanguinalis*, *Calystegia sepium*, *Rumex crispus*, *Pteridium tauricum*, *Conyza canadensis*, etc., are the main weeds of these tea plantations. Weeds of the above-mentioned group include a number of annuals with the flowering period in spring, such as *Vicia tetrasperma*, *V. angustifolia*, *Vulpia myuros*, *Bromus japonicus*, *Stellaria media*, etc.

Segetal flora in the areas under tung trees, laurels and citrus fruits resembles the tea plantations. *Sorghum halepense*, *Echinochloa crus-galli*, *Setaria glauca*, *S. viridis*, *Cichorium intybus*, etc., frequently occur on these plantations.

10.1.3 Orchards and Vineyards

Viniculture and horticulture are the leading branches of agriculture of Georgia. Large orchards with various fruit trees and vineyards have spread all over the country. Their diverse flora counts over 400 species. Common segetals can be manifested by *Cynodon dactylon*, *Setaria glauca*, *Convolvulus arvensis*, *Lavatera thuringiaca*, *Tragopogon graminifolius*, *Urtica dioica*, *Malva sylvestris*, etc.

10.1.4 Vegetable Gardens

This group is involves not only common segetal weeds but also plants growing on rich, fertilised soils. Examples of the above-mentioned plants are *Soncus asper*, *S. oleraceus*, *Chenopodium album*, *Amaranthus retroflexus*, *A. lividus*, *Portulaca oleracea*, *Cynodon dactylon*, etc.

10.1.5 Rice Fields

Most of weeds are hygrophytes were hydrophytes in the rice fields of the West Georgia in the past times. High presence of adventive weeds native to South-Eastern Asia and Mediterranean were a peculiar feature of the segetal flora of rice fields. At present, rice is not cultivated in Georgia.

The following noxious weeds occurring in the rice fields should be mentioned: *Echinochloa crus-galli*, *Cyperus difformis* and *Cyperus glomeratus*.

10.2 Ruderal Vegetation

Communities enclosed in the group of ruderal vegetation differ by their habitat and composition.

Roadside communities are formed by plants resistant to trampling down. Common plants of roadsides include *Cynodon dactylon*, *Sclerochloa dura*, *Eleusine indica*, *Eragrostis pilosa*, *E. minor*, etc.

The following species occur on soils rich in organics: *Solanum persicum*, *Hyoscyamus niger*, *Conium maculatum*, *Datura stramonium*, *Solanum nigrum*, *Phytolacca americana*, etc.

It must be noted that these communities, along with ruderals include also weeds growing in vegetable gardens, such as the species of *Amaranthus*, *Conyza canadensis*, *Ambrosia artemisiifolia*, *Xanthium strumarium* and *Anchusa italica*.

Many plants recognised as ruderals are used in medicine (e.g. *Datura stramonium*, *Hyoscyamus niger*, etc.).

10.3 Vegetation of Pastures

Pastures, whether they consist of natural grasslands, hay meadows, which have been used for grazing, or the land, which is too poor or otherwise unfit for anything but grazing, frequently include certain weeds associated with them. Many of the commonest pasture weeds are unpalatable, distasteful or even poisonous to stock plants. Others have spines, thorns or modifications unfavoured by stock.

In Georgia, summer pastures are richer in weeds than winter ones. Majority of pasture weeds are not dominants of communities, but in overgrazed areas some of them form independent communities, such as *Veratrum lobelianum* and *Rumex alpinus*.

The following plants frequently occur on lowland semi-desert pastures: *Adonis aestivalis*, *Cynanchum acutum*, *Hirschfeldia incana*, *Xanthium strumarium*, etc.

Summer pastures of subalpine belts are characterized by the presence of *Cirsium arvense*, *C. incanum*, *Veratrum lobelianum*, *Daphne glomerata*, *Ranunculus caucasicus*, *R. oreophilus*, etc. In subalpine belt poisonous *Aconitum nasutum* and *A. orientale* are also to be found.

11 Cultivated Flora

Formation of cultivated flora is constitutionally interconnected with human activities. It makes an integral part of the material culture of mankind. The most ancient memorials of human activities date back to the Palaeolithic. The earliest traces of agriculture in Georgia belong to the Mesolithic period. Among the monuments of this era are stone tools – pestles and graters. Archaeological findings of charred remains of wheat, barley, millet and flax seeds, as well as of agricultural utensils such as grain-grates, sickles, etc., confirm the diversity of the cultivated flora of Neolithic and particularly of the Eneolithic period.

Eneolithic culture of Georgia, which can be traced back to the third millennium B.C., was mainly represented by agriculture and cattle-breeding. Fruit gathering was also widely practised. Among the remains of the late Eneolithic and early Bronze Ages, such fruit of wild plants as chestnut (*Castanea*), beech (*Fagus*), oak (*Quercus*) and hazelnut (*Corylus*) have been discovered side by side with cultivated plants in the burial mounds of West Georgia.

Agriculture of ancient Georgia is best illustrated by the monuments of the early Bronze Age (second millennium B.C.). Many kinds of wheat were domesticated during that period; barley millet, flax and grapevine were subject to cultivation. In the late Bronze Era (10–7 centuries B.C.), in the States of Urartu and Diaokhi in Transcaucasia almost all branches of agriculture were well developed; cultivation of cereals, seed and forage legumes, fruit trees, grapevine and fodder-grass (especially for horse food and cattle-breeding). By that time irrigation had already been well organized.

Cultivation of many plants in ancient Georgia was mentioned in the works of writers of that time. Xenophon (the fifth century B.C.) speaks of two kinds of wheat cultivated in Colchis.

Wheat is one of the oldest domesticated plants in the world. Archaeological discoveries prove that Karthvelis (Georgians) started to cultivate wheat more than 5,000 years ago. During these millennia Karthveli farmers created many species and varieties of wheat.

The initial roots of some wheat species dating back to the Eneolith and Bronze eras (*Triticum macha*, *T. paleocolchicum*) are being preserved in the *ex situ*

G. Nakhutsrishvili, *The Vegetation of Georgia (South Caucasus)*, Geobotany Studies, DOI 10.1007/978-3-642-29915-5_11, © Springer-Verlag Berlin Heidelberg 2013

collections of present-day Georgia. Such species of wheat as the so-called Zanduri, incomparable among world wheats for their qualities and immunity against diseases, are the products of labour of Kartveli farmers. So far, modern selection has failed to breed a kind of wheat with the properties of Zanduri.

By means of phylogenetic research, a basic variety of wheat species in the agriculture of Georgia has been revealed. Some of these species bear evolutionarily close affinity to wild wheat species, or have retained some of their features. The wheats Makha (*T. macha*), Colchic spelt (*T. paleocolchicum*), Zanduri (*T. timopheevi*, *T. zhukowskyi*) are unique plants, which are still being cultivated and which still retain the basic features of wild wheat.

Because of very brittle ears, harvesting of such species as Makha and Colchic spelt proceeds in two stages: first by means of two loosely bound sticks (so called shankvi) you throw the ears into baskets and then cut the straw. The species Makha is the living relict of primary agriculture, which has been cultivated by man since the prehistoric era (Eneolith-Bronze). Another unique monument of the prehistoric culture Zanduri wheat species. Species generating Zanduri population have nothing to do either with wild or cultivated species of Asia Minor, though they reveal genetic affinity with the species growing on the modern territories of Armenia and Azerbaijan.

On the whole the following cultivated species of wheat have been recorded in the fields of Georgia: *T. monococcum* (monograin), *T. dicoccum*, *T. timopheevi*, *T. zhukowskyi*, *T. paleocolchicum* (all species), *T. durum*, *T. carthlicum*, *T. turgidum*, *T. polonicum* (all hard wheats) and *T. aestivum*, *T. compactum*, *T. macha* (all soft wheats).

Georgian endemics are: *T. paleocolchicum*, *T. timopheevi*, *T. zhukowskyi*, *T. macha* and *T. carthlicum*. Only *T. carthlicum* has penetrated into the agriculture of neighbouring countries.

At present, maize, wheat, and barley are the principal cereals in Georgia. The first one, maize, absolutely predominates in West Georgia, while the second one, wheat, is mainly cultivated in the fields of East Georgia and in small quantities is available in foothills and mountain districts (Javakheti, Kazbegi), the greater part of acreage is hold under barley (about 85 %). Winter wheat crops are encountered from the littoral or Black Sea in the West (near the town of Gagra) to the high-mountain agricultural zone (up to 2,000 m s.m.). Spring wheat crops are spread everywhere, but they prevail over winter crops only above 1,500–1,800 m.

The most widespread (and therefore the most important one at present) are three species: (1) soft wheat (*T. aestivum*), (2) Karthlian wheat (*T. carthlicum*) and (3) hard wheat (*T. durum*). The rest of the species occupy small areas.

Soft wheats are the leaders in the cereal economy of the country. Many centuries of popular selection resulted in a great number of local varieties, some of which (e.g. Dolis-puri) have retained their considerable part in national economy.

Karthlian wheat Dika (*T. carthlicum*) mainly occurs in the mountain agricultural belt within 1,000–2,300 m s.m. Popular varieties are sown, and partly those of modern pureline section. Hard wheat crops extend in the lower agricultural belt from 800 to 900 m. Its acreage can be found mainly in East Georgia.

On the Black Sea coast (in the vicinity of Gagra) small tracts of so-called English wheat (*T. turgidum*) have been preserved until now.

At present, production of wheat crops is mainly represented by selection and selectionally improved varieties, and more seldom by more popular selection varieties.

On the basis of phylogenetical research, it was ascertained that the principal specific variety of wheat has been preserved only in the agriculture of Georgia and has never even surpassed the boundaries of the land of wheat crop originators, so that primary species of crops close to wild species are known only here and used only by Georgian people.

The idea is that the creation of primary species of cultivated plants is genetically connected with the natural conditions of some geographical regions, where only useful plants of the wild flora make primordial source of popular selection. It has already been stated that Asia Anterior–South-West Asia (including Transcaucasia) is the only region of habitation of wild wheat. Therefore, it is natural to presume that the cultivation of wheat is the result of creative activities of civilized people of Asia Anterior–South-West Asia. Archaeological materials and botanical-genetic analysis of modern wheat composition give enough grounds to attribute wheat cultivation to Khetto-Iberian group of people.

Natural conditions for wheat domestication did exist only in that region.

All the aforesaid underline the primordial character of a Georgian wheat cultivation. The Georgian nidus by origin and formation of wheat species diversity is a consequence of a long process of popular selection, the main roots of which are genetically connected with the pre-historic activity of Hetto-Sumeret peoples, living in the territory of Asia-Anterior – the region of habitats of wild wheat.

The creative process of Georgian popular selection is not limited by the abovestated. The collective work of Georgian cultivators manifests itself by the cultivation of grapevine. At present, there are over 400 varieties with Georgian appellations. Versatility and antiquity of the culture is confirmed not only by the existence of numerous sorts, but by data of Eneolithic culture as well. In the times of the existence of the state of Urartu, Colchis and Diaokhi, wine-making as well as viticulture was highly developed. Georgian popular grape sorts – Saperavi, Rkatzhiteli, etc. – enjoy on our days the same unsurpassed fame, which in its turn points to the long process of selection of high-grade grape varieties.

Georgia is likewise the home of many fruit plants. Many species of wild fruit-trees (*Malus*, *Pyrus*, *Prunus*) have undoubtedly served as the principle components in the origin of cultured plants like apple, pear, and plum-trees.

It has been proved by the investigation of Georgian scientists that the Georgian apple sort Khomarduli is the initial variety for the so-called English and French paradise-apples widely diffused in European countries.

Ancient agricultural crops were formed exclusively on the basis of the utilization and cognition of aboriginal useful wild plants. In this way, primary independent nidi of agriculture arose. In the subsequent period of the cultural development and intercourse of peoples those primary agricultural groups were mutually enriched with imported plants. It has been ascertained from historical sources that in ancient

Georgia the introduction of useful foreign plants was organized on a large scale. It is obvious that the presence of the *Citrus* type of agriculture production created in Georgia from representatives of foreign subtropical plants indicates a great interchange carried on between the peoples inhabiting lands with different ecogeographical conditions. Many of these plants, no doubt passed through a number of "replantation" points to the way from their initial habitats to their new home in Georgia. These points played a certain part in acclimatization and selection.

Gardens of royal and feudal families of ancient Georgia played a major role in the introduction of foreign plants, in particular Citrus, decorative and other useful ones.

At present, numerous representatives of foreign vegetation – the tea-bush, the tangerine-tree, subtropical fruit-trees and technical and decorative plants, etc., are important (and in some regions the chief) objects of agricultural production (Menabde 1961).

12 Georgia: Protected Areas

Georgia founded the first strict nature reserve in the Caucasus Ecoregion, Lagodekhi Strict Nature Reserve, in 1912 to conserve pristine Caucasus broad-leaved forests and associated wildlife. At the end of the 1990s (Soviet era) the Protected Area system in Georgia included 15 Strict Nature Reserves only, which covered 2.4 % of country's territory – mostly forests of different types. After the collapse of Soviet Union the system, with support of donors, such as GEF/World Bank, German Government, Government of Norway, and with technical assistance of WWF and some other international and national organizations, has been developing rapidly in both directions: extension and transformation-diversification. Today Georgia has 14 State (Strict) Nature Reserves (IUCN Category I), 9 National Parks (IUCN II), 14 Natural Monuments (IUCN III), 16 Sanctuaries (IUCN IV), 2 Protected Landscapes (IUCN V) and 5 Multiple Use Areas (IUCN VI) covering a total area of around 520,000 ha or 7.5 % of the country's territory (three times more than in the 1990s) (Fig. 12.1). The system protects around 10 % of the country's forest area, among them: pristine forests Colchic humid broad-leaved, mixed and swamp forests (e.g. Kintrishi Strict Nature Reserve, Mtirala and Kolkheti National Parks), East Caucasus temperate broad-leaved forests (e.g. Lagodekhi and Tbilisi Strict Nature Reserves), high mountain birch and pine forests (e.g. Kazbegi and Tusheti National Parks), dry open pistachio-juniper woodlands (Vashlovani National Park), East Caucasus flood-plain forests (e.g. Gardabani and Chachuna Sanctuaries), etc. High mountain grasslands are main object for protection of newly established Javakheti National Park. However, Georgia's protected areas system still needs improvement and development. Some new protected areas, effectively functional buffer zones, as well as corridors between protected areas need to be established to allow animal migrations and certain threatened ecosystems need to be set aside. Principal fauna species are globally threatened: West Caucasus Tur (wild goat, endemic for the Greater Caucasus – *Capra caucasica*), Bezoar goat (*C. aegagrus*), Caucasian salamander (relict and endemic species for the West

G. Nakhutsrishvili, *The Vegetation of Georgia (South Caucasus)*, Geobotany Studies, DOI 10.1007/978-3-642-29915-5_12, © Springer-Verlag Berlin Heidelberg 2013

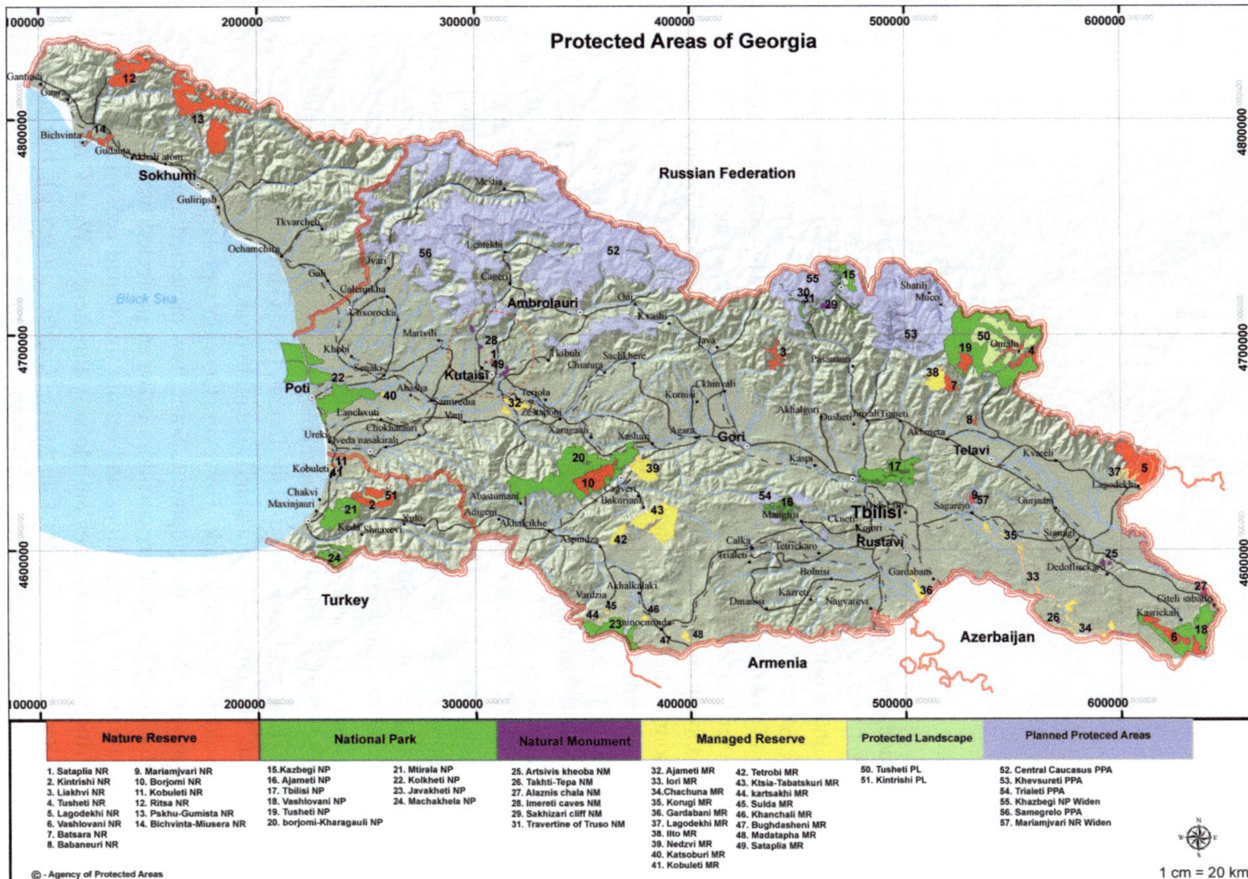

Fig. 12.1 Protected areas of Georgia (from the Agency of Protected Areas of Georgia)

Lesser Caucasus), as well as brown bear (*Ursus arctos*), European lynx (*Lynx lynx*), East Caucasus Tur (endemic – *Capra cylindricornis*), C. Caucasian red deer (*Cervus elaphus maral*), etc. Few individuals of globally endangered Caucasian leopard (*Panthera pardus ciscaucasica* = *P.p. saxicolor*) occur in Tusheti and Vashlovani National Parks (Williams et al. 2006; IUCN 2009; Zazanashvili et al. 2009).

Zusammenfassung

Die Mannigfaltigkeit der physisch-geographischen Faktoren: die komplizierte Gliederung des Reliefs, die kontrastierenden Höhenstufen wie auch die Lage Georgiens an der Grenze von biogeographischen Groβräumen verschiedener Genese bestimmen den Reichtum und die Vielfalt der Pflanzenwelt Georgiens. Auf einem verhältnismäβig kleinen Territorium entwickelten sich die verschiedenartigsten Pflanzengesselschaften, angefangen von Halbwüsten in den Ebenen und Hügeln im Osten über üppige Wälder unter feuchtem, fast subtriposchem Klima im Vorgebirgsland der Kolchis bis zu der eigentümlichen kryophilen Hochgebirgsvegetation des Kleinen Kaukasus im Südwesten und insbesondere des Groβen Kaukasus im Norden.

Die Gliederung des Reliefs und die Struktur der Gebirge bewirken ein hohes Maβ an geographischer und ökologischer Isolation der regionalen Ökosysteme, was im hohen Grad des Endemismus (ungefähr 1500 Arten) seinen Niederschlag findet. Auf dem Territorium Georgiens trifft man gegen 5000 Arten wildwachsender und verwilderter Gefäβpflanzen.

Im Eozän unterschied sich die Flora sehr wesentlich von der gegenwärtigen. Die gröβte Ähnlichkeit mit jener hat heute aller Wahrscheinlichkeit nach die indisch-australische Flora mit ihren immergrünen Baumarten, Baumfarnen, Palmen u.a. Es wird angenommen, daβ die waldlosen Formationen der trockenresistenten Vegetation im Kaukasus bereits im Miozän weit verbreitet waren. Bereits im Anfang des Pliozäns entsteht auf jenem Teil des Festlandes, wo jetzt Westgeorgien und einige angrenzende Territorien des Schwarzmeergebietes liegen, das kolchische Refugium mit Relikten einer älteren mesophilen Waldflora, begünstigt durch das dort herrschende verhältnismäβig warme und feuchte Klima. Das kolchische Refugium hat bei der Erhaltung reliktischer Arten im Lauf des Pliozäns eine groβe Rolle gespielt, und zwar besonders im Laufe des Quartärs einschlieβlich der Epochen der groβen Verweisungen.

In Westgeorgien fehlt vollständig der waldlosen Gürtel der Trockenvegetation, vielmehr sind Täler und Berghänge schon vom Meerespiegel an mit Wäldern bedeckt. Deshalb ist die vertikale Stufung (Gürtelung) einfacher und durch vier Grundstufen darzustellen: Waldstufe (von der Küste bis zu 1900 m s.m.), subalpine (1900–2400 m), alpine (2400–2900 m), subnivale und nivale (2900 m) Stufe.

Im Ostgeorgien ist die vertikale Gürtelung komplizierter. Hier sind 6 Hauptstufen (Gürtel) ausgeprägt: der Gürtel der Halbwüsten, der trockenen Steppen und

G. Nakhutsrishvili, *The Vegetation of Georgia (South Caucasus)*, Geobotany Studies, DOI 10.1007/978-3-642-29915-5, © Springer-Verlag Berlin Heidelberg 2013

der trockenen Lichtwaldes (150–600 m s.m.), der Waldgürtel (600–1900 m), der subalpine (1900–2500 m), der alpine (2500–3000 m), der subnivale (3000–3600 m) und der nivale (3600 m s.m.) Gürtel.

Innerhalb der Waldstufe und der subalpinen Stufe der südgeorgischen Hochlandes (Dschavacheti) entwickeln sich auch waldlose Vegetationstypen: Bergsteppen und Trockenwiesen.

References

Aeschimann D, Lauber K, Moser D, Theurillar JP (2004) Flora alpina B.2. Haupt, Bern Stuttgart, Wien

Abdaladze OG, Kikvidze Z (1991) Canopy impact on CO_2 assimilation in *Trifolium ambiguum* M. B. in the aubalpine belt of the central Caucasus. Ecology 1: 33–37

Abdaladze OG (1994) CO_2-gas exchange in high mountain plants of Georgia (Central Greater and Lesser Caucasus). A short review of the doctoral thesis. Tbilisi

Akhalkatsi M, Abdaladze O, Smith WK, Nakhutsrishvili G (2004a) Mechanisms of treeline stability in the Caucasus Mountains of Georgia: low-temperature photoinhibition and interspecific facilitation.In: Abstracts of scientific Meeting "Botany 2004", Alpine diversity: adapted to the Peaks Snowbird, Utah: 55, July 31–Aug 5 2004

Akhalkatsi M, Abdaladze O, Nakhutsrishvili G, Smith WK (2004b) Global warming effect on birch seedling establishment at tree line in the Central Caucasus. Botanikertagung, Braunschweig: 386, 05 bis 10 Sept 2004

Akhalkatsi M, Abdaladze O, Nakhutsrishvili G, Smith WK (2006) *Rhododendron caucasicum* and microtopography extend the *Betula litwinowii* alpine treeline (Caucasus Mountains, Georgia). Arct Antarct Alp Res 38(4):481–488

Aleksandrova VD (1983) Vegetation of polar deserts of the USSR. Nauka, Leningrad

Bedoshvili DO (1985) Groups of associated species in the plant communities of the alpin meadows in the Kazbegi region (Central Caucasus). Bot Zh 70:1523–1528

Bohn U, Golub G, Hettwer Ch. (eds) 2003. Karte der naturlichen Vegetation Europas. - Map of the natural vegetation of Europe. Legende, Legend. Bundesamt fur Naturschutz, Bonn.

Borisova IV (1972) Seasonal dynamics of a plant community. In: Lavrenko EM, Korchagin AA (eds) Polevaya Geobotanika. Nauka, Leningrad

Box E, Fujiwara K, Nakhutsrishvili G, Zazanashvili N, Liebermann R, Miyawaki A (2000) Vegetation and landscapes of Georgia (Caucasus), as a basis for landscape restoration. Bull Inst Environ Sci Technol Yokohama Natn Univ 26:69–102

Braun-Blanquet J (1964) Pflanzensoziologie. Grundzüge der Vegetationskunde, 3rd ed., Springer, Wien

Burga CA, Klötzli F, Miehe G (2004) Waldgrenze: Phänomen und globaler Vergleich. In: Burga CA, Klötzli F, Grabherr G (eds) Gebirge der Erde: Landschaft, Klima. Eugen Ulmer, Stuttgart, pp 37–44

Bush NA (1935) Botanical and geographical studies of the Caucasus. Izd AN SSSR, Moscow/ Leningrad

Caldwell M (1968) Solar ultraviolet radiation as an ecological factor for alpin plants. Ecol Monogr 38:243–268

Caldwell MM (1989) The changing solar ultraviolet climate and ecological consequences for higher plants. Trends Ecol Evol 4(12):363–367

Cernusca A (1976) Bestandesstructur, Mikroklima und Energiehaushalt von Pflanzenbestanden des alpinen Grasheidegürtels in den Hohen Tauern. In: Cernusca A (ed) Alpine Grasheide

G. Nakhutsrishvili, *The Vegetation of Georgia (South Caucasus)*, Geobotany Studies, 229
DOI 10.1007/978-3-642-29915-5, © Springer-Verlag Berlin Heidelberg 2013

Hohe Tauern. Veröff. Österr. MaB-Hochgebirgsprogramms Hohe Tauern 1:4. Wagner, Innsbruck, pp 25–45

Dmitrieva AA (1959) Key to Plants of Adjara. Publishing House of the Georgian SSR Academy of Sciences, Tbilisi

Dolukhanov AG (1946) Upper limits of alpine vegetation in the headwaters of Avar Koy-Su (Dagestan). Trudy Tbilissk Bot Inst 9:131–154

Dolukhanov AG (1956) Natural regeneration of beech in the main types of beech forests of Georgia. Trudy Tbilissk Bot Inst 18:69–114

Dolukhanov AG (1966) Reguliarities of geographical diversity of vegetation and upper timberline in the Trans caucasian mountains. Probl Bot 8:196–207

Dolukhanov AG (1969) Flora and vegetation of subnival landscapes of the Didi Liakhvi River and the Keli uplan (Greater Caucasus). Bot Zh 54:1662–1674

Dolukhanov AG (1978) Timberline and subalpine belt in Caucasus Mountains, USSR. Arct Alp Res 10(2):409–422

Dolukhanov AG (1980) Colchic underwood. Metsniereba, Tbilisi

Dolukhanov AG (2010) Forest vegetation of Georgia. Universal, Tbilisi

Dolukhanov AG, Sakhokia MP, Kharadze AL (1942) On high mountain belts of the Caucasus. Trudy Tbilissk Bot Inst 8:113–135

Dolukhanov AG, Sakhokia MP, Kharadze AL (1946) Main features of the vegetation cover of Upper Svanetia. Trudy Tbilissk Bot Inst 9:79–130

Dolukhanov AG, Urushadze TF (1968) On the nature of litter types of Caucasian beech forests. Lesovedenie 4:3–13

Doluchanov G, Nachucrišvili G (2003) Die natürlichen Vegetationsformationen Europas und ihre Untergliederung. Hygrophile thermophytische Laubmischwälder. In: Bohn U, Gollub G, Hettwer C (eds) Karte der natürlichen Vegetation Europas. Map of the Natural Vegetation of Europe, Maßshtab/Scale 1:2 500 000 m. Federal Agency for Nature Conservation, Bonn, pp 384–388

Ellenberg H (1996) Vegetation Mitteleuropas mit den Alpen. Ulmer Verlag, Stuttgart

Erschbamer B, Mallaun M, Unterluggauer P, Abdaladze O, Akhalkatsi M, Nakhutsrishvili G (2010) Plant diversity along altitutinal gradients in the Central Alps (South Tyrol, Italy) and in the Central Greater Caucasus (Kazbegi region, Georgia). Tuexenia 30:11–29

Franz H (1979) Ökologie der Hochgebirge. Ulmer Verlag, Stuttgart

Feodorov FF (1952) The history of high mountain flora of the Caucasus in the quaternary period as an example of autochthonous development of the Tertiary floristic basis. In: Kristophovich AN (ed) Materialy po chetvertichnomu periodu. SSSR, 3, Moscow, pp 49–86

Gagnidze RI (1974) Botanical and geographical analysis of the florocoenotic complex of tall herbaceous vegetation of the Caucasus. Metsniereba, Tbilisi

Gagnidze RI (1977) The ecological-coenotic characteristic and analysis of vertical distribution of tall herbaceous species in the Caucasus. Zametki Sistem Geogr Rastenii Inst Bot AN GSSR 34:41–83

Gagnidze RI (2000) Diversity of Georgia's flora. In: Berutchashvili N, Kushlin A, Zazanashvili N (eds) Biological and landscape diversity of Georgia. WWF Georgia Country Office, Tbilisi, pp 21–33

Gamtsemlidze ZG (1977) Bioecological and coenotic peculiarities of Kazbegi subnival plants. In: Nakhutsrishvili G (ed) The high mountain ecosystems of Kazbegi. Metsniereba, Tbilisi, pp 44–49

Gamtsemlidze ZG (1979) On bud of regeneration of subnival plants of Kazbegi (Central Caucasus). Izv AN GSSR, Serya Biologicheskaya, Tbilisi 5:432–450

Geideman TS (1932) The studies on vegetation of high mountain pastures of Kurdistan. Trudy po geobot obsledov pastbish SSR Azerbaidzhana, Publishing House of the Academy of Sciences of Azerbaijan SSR, Baku, ser. B, 5

Gerasimov IP (ed) (1966) The Caucasus. Nauka, Moscow

Golitsin SV (1939) Skeriani – shrub phytocoenoses of humid forested mountains of Adjaria. Trudy Voronezhsk Gos Universiteta XI 2:141–175
Gorishina TK (1979) Plant ecology. Nauka, Moscow
Gorishina TK, Nakhutsrishvili GS, Khetsuriani LD (1985) On some anatomo-physiological peculiarities of photosynthetic organs of high mountain plants of the Central Caucasus. Ökologia 5:7–14
Grebenshikov O (1965) Geobotany Dictionary Russian-English-German-French. Nauka, Moscow
Grabherr G, Mucina L (eds) (1993) Die Pflanzengessellschaften Österreichs. Band 2. Fischer, Jena
Grabherr G (1997) The high Mountain ecosystems of the Alps. In: Wielgolaski FE (ed) Ecosystems of the World 3: Polar and Alpine Tundra. Elsevier, Amsterdam, pp 97–121
Grossheim AA (1936) The analysis of Caucasian flora. Trudy Bot ins Azerb fil AN SSSR. Izd Az Fil AN SSSR, Baku
Grossheim AA (1946) The tasks of flristic investigation of the Caucasus (a degree of floristic understanding of the Caucasus). Sovetskaia botanika 2:116–121
Grossheim AA (1948) Vegetational cover of the Caucasus. MOIP, Moscow
Gulisashvili V (1949) On some specificities of primary forests of *Fagus orientalis* in East Georgia. Trudi Instituta Lesa Akad. Nauk Gruzii 1: 4–20
Gulisashvili VZ (1964) Natural zones and natural history areas of the Caucasus. Nauka, Moscow, p 326
Steinacher G, Wagner J (2010a) Flover longevity and duration of pistil receptivity in high mountain plants. Flora 205(6):376–387
Holzner W (ed) (1989) Biotoptypen in Österreich. Umweltbundesamt, Wien
IUCN (2009) IUCN Red List of threatened species. Version 2009. http://www.iucnredlist.org. Accessed 14 Sept 2009
Ivanishvili MA (1973) Flora of thorn astragal tragacanth formation of the northern slope of the Greater Caucasus. Metsniereba, Tbilisi
Ivanishvili MA (1998) The thorn-cushion vegetation in the Caucasus. In: Nakhutsrishvili G, Abdaladze O (eds) Plant life in High-Mountains. Diogen, Tbilisi, pp 43–50
Janelidze C, Margalitadze N (1977) History of Forest Vegetation in Holocene. In: Nakhutsrishvili G (ed) Alpine Ecosystems of Kazbegi. Publishing House of the Georgian Academy of Sciences, Tbilisi, pp 17–21
Jaroshenko PD (1942) On vegetation changes in the forested region of the Trans-Caucasus. Izd Armyansk Fil AN SSSR, Erevan 7(21):32–46
Kakhadze JR (1947) The Georgia in the Jurassic time. Bull Geolog Inst AN Gruzii 8, Tbilisi
Ketskhoveli NN (1935) The vegetation of Georgia. Publishing House of Tbilisi State University, Tbilisi
Ketskhoveli NN (1960) Vegetational cover of Georgia. Metsniereba, Tbilisi (in Georgian)
Ketskhoveli N, Kharadze A, Gagnidze R (eds). 1971–2011. Flora of Georgia, 2nd ed. Vols. 1–16. Metsniereba, Tbilisi (in Georgian)
Kharadze AL (1948) On periglacial vegetation of the Caucasus. Bull Georg Acad Sci 9–10:615–622
Kharadze AL (1960) An endemic hemixerophilous element of Greater Caucasus uplands. Probl Bot 5:115–126
Kharadze AL (1965) On the subnival belt of the Greater Caucasus. Zametki Sistem Geogr Rastenii Inst Bot AN GSSR 25:103–104
Kharazishvili D (2005) Threeline Vegetation of the River Chirukhistskali(South Colchis, Adjara). Proc Georgian Acad Sci 1(3):32–43
Khintibidze L (1990) Xerophilous floristic complexes of South Georgia. Manuscript of the Doctor of Science dissertation. Tbilisi. (in Russian)
Kikvidze Z, Nakhutsrishvili G (1998) Facilitation in the subnival vegetation patches. J Veg Sci 9:222–226
Kimeridze KR (1965) Variegated fescues of Kavkasioni. Metsniereba, Tbilisi (in Georgian)
Kimeridze KR (1992) Bog vegetation of Georgia. Tbilisi (unpublished manuscript)

Kolakovsky AA (1961) Plant world of Colchis. Publishing House of Moscow University, Moscow
Kolakovsky AA (1964) Pliocene flora of the Kodori River. Metsniereba, Sukhumi
Kolakovsky AA (1974) Vertical belts of Colchis forest vegetation in the Tertiary period. Trudy Tbilissk Inst Lesa 21:98–116
Kolakovsky AA (1980–1986) Flora of Abkhazia. Vols. 1–4. Metsniereba, Tbilisi
Kolakovsky AA, Shakril AK (1976) Sarmatian floras of Abkhazia. Trudy Sukhumsk Bot Sada 22:98–148
Kolakovsky AA, Shakril AK (1978) Kimmerian flora of Gul'ripsha. Trudy Sukhumsk Bot Sada 24:134–156
Körner Ch (1998) Are-assessment of high elevation treeline positions and their explanation. Öcologia 115:45–459
Körner Ch (2003) Alpine plant life. Functional plant ecology of high mountain ecosystems. Springer, Berlin
Korner Ch (2012) Alpine Treelines. Functional ecology of the global high elevation tree limits. Springer, Basel.
Körner Ch, Cernusca A (1988) A semiautomatic, recording diffusion porometer and its performance under alpine field conditions. Photosynthetica 10:172–181
Körner Ch, Larcher W (1988) Plant life in cold climates. In: Long S, Woodward FI (eds) Plant and temperature. CUP, Cambridge, pp 25–27
Körner Ch, Mayr R (1980) Stomatal behaviour in alpine plant communities between 600 and 2600 m plants and their atmospheric environment. In: Woodward FI (ed) 21 symposium of British ecological society. Edinburgh, 1979. Blackwell, Oxford, pp 124–131
Körner Ch, Paulsen J, Spehn EM (2011) A definition of mountains and their bioclimatic belts for global comparisons of biodivrsity data. Alp Bot 121:73–78.
Korotkov KO (1986) Vegetation of the Gouton Mountain East Caucasus. In: Kamelin RV (ed) Vegetation of high altitutes. Nauka, Moskow, pp 127–130
Kreeb K (1974) Ökophysiologie der Pflanzen. Fischer, Jena
Kvachakhidze RK (1979) High mountain forests of the southern slope of the Greater Caucasus and main directions of their changes (within Georgian SSR). Metsniereba, Tbilisi
Kvachakidze RK (2009) Vegetation of Georgia. Publishing House of the Georgian National Academy of Sciences, Tbilisi (In Georgian)
Lachashvili I, Lachashvili N, Khachidze M (2004) Synopsis of flora of the Vashlovani State Reserve. Universal, Tbilisi
Lachashvili N, Khachidze M (2009) Typology and distribution patterns of the vegetation of Kiziki (East Georgia). *kavkazskiy geographicheskiy jhurnal* 9-90-96. (in Russian)
Lachashvili N, Khachidze M (2010) Desert flora and vegetation of Georgia. Universal, Tbilisi.
Ladinig U, Wagner J (2009) Dinamics of flower development and Vegetative shoot growt in the high mountain plant Sxifraga bryoides L. Flora 204:63–73
Larcher W (1975) Pflanzenökologische Beobachtungen in der Paramostufe der venezolanischen Anden. Anz math naturwiss Kl sterr Akad Wiss 11:2–20
Larcher W (1977) Ergebnisse der IBP Projekts"Zwergstrauchheide Patscherkofel". Sitzungsberichte Osterr Akad Wiss(Wien) Math Naturwiss KL Abt 1186–301
Larcher W (1980) Klimastress im Gebirge: Adaptationstraining und Selektionsfilter für Pflanzen. Rheinisch-Westfal Akad Wiss Leverkusen: Westdeutsche Verlag 291:49–79
Larcher W (1983) Ökophysiologische Konstitutionseigenschaften von Gebirgspflanzen. Berl Deutsch Bot Ges 96:73–85
Larcher W (1987) Stress bei Pflanzen. Naturwiss 74:158–167
Larcher W, Nakhutsrishvili GS (1982) On pheno- and throphorhythmotypes of high mountain plants. Ekologia 4:13–20
Larcher W, Kainmuller C, Wagner J (2010) Survival types of high mountain plants under extreme temperatures. Flora 205(1):3–18
Lavrenko EM (1958) Concerning position of the forest vegetation of the Caucasus in the system of botanical-geographic division of the Palearctic. Bot Zh 43(9):1237–1253

Magakian AK (1941) Vegetation of Armenian SSR. Izd AN SSSR, Leningrad
Menabde VJ (1961) Cultivated flora of Georgia. In: Sakhokia MF (ed) Botanical excursions over Georgia. Publishing House of the Academy of Sciences of Georgian SSR, Tbilisi, pp 69–76
Moser W (1978) Neues von der botanischen Forschungsstation "Hoher Nebelkogel". Jahrb Verein zum Schutze der Alpenpflanze und Tiere 33:125–133
Nagy L, Grabherr G (2009) The biology of Alpine habitats. OUP, Oxford
Nakhutsrishvili Sh (1966) Main features of vegetational cover of the southern upland of Georgia. Probl Bot 8:106–111
Nakhutsrishvili GS (1974) Ecology of high mountain plants and phytocenoses of the Central Caucasus. Metsniereba, Tbilisi
Nakhutsrishvili SH (1981) Peculiarities of the structure and development rhythm of high mountain plants. In: Serebryakova T (ed) Life forms: structure, spectra and evolution. Nauka, Moscow, pp 249–264
Nakhutsrishvili GS (1988) The ecological characteristic of high mountain meadow vegetation on the example of Kazbegi (Central Caucasus). Ecology of high mountains. Metsniereba, Tbilisi, pp 175–200
Nakhutsrishvili G (1998a) Hochgebirgsvegetation Georgiens (Kaukasus). In: Nakhutsrishvili G, Abdaladze O (eds) Plant life in High-Mountains. Diogen, Tbilisi, pp 93–100
Nakhutsrishvili G (1998b) The vegetation of the subnival belt of the Caucasus mountains. Arct Alp Res 30:222–226
Nakhutsrishvili G (1999) The vegetation of Georgia (Caucasus), vol 15, Braun-Blanquetia. Dipartimento di Botanica ed Ecologia dell'Università, Camerino, pp 5–74
Nakhutsrishvili G (2003) High Mountain vegetation of the Caucasus region. In: Nagy L, Grabherr G, Körner C, Thompson DBA (eds) Alpine biodiversity in Europe. Springer, Berlin/Heidelberg, pp 93–103
Nakhutsrishvili GS, Gamkrelidze TA, Gamtsemlidze ZG, Sanadiradze GM (1990) Zur Frage der Überlebensstrategien von Pflancen in der subnivalen Stufe des Kaukasus. Giorn Bot Ital 124:781–791
Nakhutsrishvili GS, Gamtsemlidze ZG (1984) Plant life in extremal environment of the high mountains. Nauka, Leningrad
Nakhutsrishvili G, Gagnidze R (1999) Die subnivale und nivale ochgebirgsvegetation des Kaukasus. Phytocoenosis 11:173–182
Nakhutsrishvili G, Ozenda P (1998) Aspect Geobotaniques de la Haute Montagne dans le Caucase. Essai de Comparaison avec les Alpes. Ecologie 29:139–144
Nakhutsrishvili G, Zazanashvili N & Batsatsashvili K (2011) Why the Colchis and Hyrcan relict forests of the western Eurasian Caucasus should be considered "temperate rainforest". – In: DellaSala D (ed.) Temperate and Boreal Rainforests of the World: Ecology and Conservation. Island Press, Wadhington DC, Pp. 214–221
Onipchenko VG, Blnnikov MS (eds) (1994) Experimental investigation of alpine plant communites in the nordwesten Caucasus. Vroff, Geobot. Institut. ETH (Stiftung Rubel), Zurich
Ozenda P (1985) La végétation de la chaîne alpine dans l'espace montagnard européen. Masson, Paris/NewYork/Barcelone/Milan/Mexico/Sao Paulo
Ozenda P (2002) Perspectives pour une géobiologie des montagnes. Biologie / Presses Polytecchniques et Universitaires Romande, Lausanne
Panjutin PS (1939) Tall herbaceous vegetation of the West Caucasus. Izvestia Russk Geogr Obschestva 71(9):27–45
Pauli H, Gottried M, Grabherr G (2011) Nemorele und meditterane Hochgebirge: Klima, Vegetationstufen, Artenfielfalt und Klimawandel am Beispiel der Alpen und Spanischen Sierra Nevada, Anhuf D, Fickert T and Gruninger F (eds) Okozonen im Wandel. Selbstverlag Fach Geographie der Universitat Passau, Passau, pp 145–158
Pedrotti F (1998) La cartographie géobotanique des biotopes du Trentin (Italie). Ecologie 29 (1–2):105–110

Prima VM (1974) Subnival flora of the East Caucasus, its composition, ecological-biological and geographical analysis. In: Galushko AJ (ed). Flora and vegetation of the East Caucasus. Publishing House of the State University of North Ossetia, Orjonikidze, pp 46–48

Reisigl H, Pitschmann H (1958) Obere Grenzen von Flora und Vegetation in der Nivalstufe der Zentralen Ötztaler Alpen (Tirol). Vegetatio 8:93–128

Rikli M (1943) Das Pflanzenkleid der Mittelmeerländer. Zweite Auflage B 1. Verlag Hans Huber, Bern

Pott R (1996) Biotoptypen: Schützenswerte Lebensräume Deutschlands und angrenzender Regionen. Ulmer, Stuttgart (Hohenheim)

Sakai A, Larcher W (1987) Frost survival of plants. Springer, Berlin, p 321

Sakhokia MF (ed) (1958) Botanical excursions over Georgia. Izd AN Gruzinskoi SSR, Tbilisi

Sakhokia MF (1980) In: Lavrenko (ed) Vegetation of the European part of the USSR (Evksinsk road-leafed forests). Nauka, Leningrad, pp 183–194

Sakhokia M (1983) Agrobotanical Review of Pastures and Hay Meadows of the Kazbegi Region. Metsniereba, Tbilisi

Sakhokia M, Khutsishvili E (1975) Conspectus Florae Plantarum Vascularium Chewi. Metsniereba, Tbilisi

Sanadiradze GM (1986) Water balance of the High-Mountain plants of the Central Caucasus. Dissertation, Tbilisi

Savile DBO (1972) Arctic adaptations in plants. Monographia 6:6–81

Semikhatova OA (1965) On respiration of high mountain plants. In: Sukhachev VN (ed) Voprosi biologii i fiziologii rastenii v usloviakh visokogorii. Probl Bot 7 pp 143–158

Serebriakov IG (1961) The rhythm of seasonal development of Siberian tundra plants. Bulleten' Moskovskogo obschestva Ispitatelei Prirodi Otd Biol 37(2):3–20

Sharashidze RV (1967) Dzelkova woods of Georgia. Avtoref Diss kand biol nauk. Metsniereba, Tbilisi

Shatilova I, Mchedlishvili N, Rukhadze L, Kvavadze E (2011) The history of the flora and vegetation of Georgia (South Caucasus). Publishing House of the Academy of Sciences of Georgia, Tbilisi

Shelyag-Sosonko UR, Didukh JP (1980) Yalta mountain-forest state reservation (botanical and geographical studies). Naukova Dumka, Kiev

Shetekauri S, Gagnidze R (2000) Diversity of high-mountain endemic flora of the Greater Caucasus. In: Berutchashvili N, Kushlin A, Zazanashvili N (eds) Biological and landscape diversity of Georgia. WWF Georgia Country Office, Tbilisi, pp 135–159

Shetekauri Sh and Jakoby M (2009) Mountain flowers & trees of Caucasia. Bunebaprinti, Istanbul.

Sokhadze EV, Sokhadze ME (1991) Steppe and semi-desert vegetation of East Georgia (unpublished manuscript)

Steinacher G, Wagner J (2010) Flover longevity and duration of pistil receptivity in high mountain plants. Flora 205(6):376–387

Steshenko AP (1967) Peculiarities of the seasonal rhythm of plants at the limiting altitudes in the Pamirs. In: Yusufbekov K, Agakhanyants O (eds) Rastitel'nost i rastenievodcheskoe osvoenie Pamira. Donish, Dushanbe, pp 170–203

Svanidze TI (1972) Jurassic flora of Georgia. Avtoref dokt diss, Tbilisi

Takhtajan AL (1946) On the history of vegetation development in Armenia. Trudy Bot Inst Akad Nauk Armenii, Erevan 4:51–107

Takhtajan AL (1978) The floristic regions of the World. Nauka, Leningrad

Tappeiner U, Cernusca A, Nakhutsrishvili GS (1989) Bestandenstruktur und Lihtklima ausgewählter Pflanzenbestande der subalpine Stufe des Zentralkaukasus. Sitzber, Osterr Akad d Wiss 5:395–421

Tchotchieva KI (1965) Flora and vegetation of Chaudini horizon of Guria. Metsniereba, Tbilisi

Tikhomirov BA (1963) Studies in plant biology of Arctic. Nauka, Moscow

Tumajanov II (1938) Forests of mountainous Tushetia. Trudy Tbilissk Bot Inst 5:105–248

Tumajanov II (1955) On the post-pliocene history of forest vegetation of the North Caucasus. Trudy Tbilissk Bot Inst 17:161–219

Tumajanov II (1980) Caucasian meadows and elfin woodlands. In: Lavrenko (ed) Vegetation of the European part of the USSR Nauka, pp 198–202

Urushadze TF (1987) Soils of mountain forests of Georgia. Metsniereba, Tbilisi

Vasilevich V, Bohn U (2003) Mesophytische und hygromesophytische Nadel und Laub-Nadelwalder. In: Bohn U, Golub G, Hettwer Ch. (eds) Karte der naturlichen Vegetation Europas. – Map of the natural vegetation of Europe. Legende, Legend. Bundesamt fur Naturschutz, Bonn, pp 178–249

Voskanyan V (1976) Flora and vegetation of the upper part of alpine and subnival belts of Mt. Aragats. Biologicheskiy zhurnal Armenii 19: 19–26.

Williams L, Zazanashvili N, Sanadiradze G, Kandaurov A (eds) (2006) Ecoregional conservation plan for the Caucasus. WWF, KFW, BMZ, CEPF, MacArthur Foundation. Printed by Signar Ltd. 220. http://assets.panda.org/downloads/ ecp_second_edition.pdf

Walter H (1968) Die Vegetation der Erde 2. Gustav Fischer Verlag, Jena/Stuttgart

Walter H (1974) Die Vegetation Osteuropas. Nord und Zentralasiens, Jena/Stuttgart

Walter H, Breckle S (1986) Okologie der Erde B 3: Spezielle Okologie der Gema_igten und Arktischen Zonen Euro-Nordasiens. Gustav Fischer Verlag, Stuttgart

Zalensky OV (1982) Ecological and physiological aspects of studies of photosynthesis and its relationship with respiration. Thesisi Doct. Diss. Leningrad

Zazanashvili N (1999) On the Colkhic vegetation. In: Klötzli F, Walther GR (eds) Conference on recent shifts in vegetation boundaries of deciduous forests, especially due to general global warming. Birkhäuser Verlag, Basel/Boston/Berlin, pp 191–197

Zazanashvili N, Manvelyan K, Askerov E, Krever V, Kalem S, Avcioglu B, Galstyan S, Mnatsekanov R, Bitsadze M (2009) Territorial protection of globally threatened species in the Caucasus. In: Zazanashvili N, Mallon D (eds) Status and protection of globally threatened species in the Caucasus. CEPF, WWF. Contour Ltd, Tbilisi, pp 222–231

MIX
Papier aus verantwortungsvollen Quellen
Paper from responsible sources
FSC® C105338

If you have any concerns about our products,
you can contact us on
ProductSafety@springernature.com

In case Publisher is established outside the EU,
the EU authorized representative is:
Springer Nature Customer Service Center GmbH
Europaplatz 3, 69115 Heidelberg, Germany

Printed by Libri Plureos GmbH
in Hamburg, Germany